Electron Microscopy and Analysis Group Conference (EMAG 2017)

Journal of Physics: Conference Series Volume 902

Manchester, United Kingdom
3 - 6 July 2017

Editor:

Caterina Ducati

ISBN: 978-1-5108-4993-8
ISSN: 1742-6588

Printed from e-media with permission by:

Curran Associates, Inc.
57 Morehouse Lane
Red Hook, NY 12571

Printed by Curran Associates, Inc. (2017)

For permission requests, please contact the Institute of Physics at the address below.

Institute of Physics
Dirac House, Temple Back
Bristol BS1 6BE UK

Phone: 44 1 17 929 7481
Fax: 44 1 17 920 0979

techtracking@iop.org

Additional copies of this publication are available from:

Curran Associates, Inc.
57 Morehouse Lane
Red Hook, NY 12571 USA
Phone: 845-758-0400
Fax: 845-758-2633
Email: curran@proceedings.com
Web: www.proceedings.com

Table of contents

Volume 902

Electron Microscopy and Analysis Group Conference 2017 (EMAG2017)
3–6 July 2017, Manchester, UK

Accepted papers received: 15 September 2017
Published online: 16 October 2017

Preface

In-situ electron microscopy

Technique Development

EM Applications

Electron Microscopy and Analysis Group Conference 2017 (EMAG2017)

Manchester, UK

3-6 July 2017

Editorial

The 2017 meeting of the Electron Microscopy and Analysis Group of the Institute of Physics was held in Manchester during the first week of July, as part of the Microscience and Microscopy Congress (mmc2017) organised by the Royal Microscopical Society. EMAG2017 hosted 87 oral presentations, including invited, contributed and flash talks. The large Exhibition in the hall of Manchester Central was the focus of interesting discussions and networking activities during the conference, complemented by busy and vibrant poster sessions.

This year we had an exceptional range of contributions, with over 220 abstracts from national and international research groups. At a time when so much is happening in terms of dynamic experiments and EM technique development, it was exciting to hear how the most recent advances are driving our understanding of materials. As in the conference, in this volume the scientific contributions are combined in broader thematic areas, which I hope will give you a flavour of the current research in Electron Microscopy.

Over the years I have followed the evolution of publishing practices in the electron microscopy community, and I have looked with some concern at the steady decay in the number of Proceedings papers relative to the number of abstracts submitted for the EMAG biennial meetings. This ratio was above 90% in 1997, and is now just 15%. Although the Proceedings continue to be peer-reviewed, are fully indexed and offered to the scientific community through Open Access, it seems that the pressure to publish novel results in higher impact journals has a strong influence on our output, possibly because of the increased cost of electron microscopy research. Interestingly, a citation analysis using the Web of Science shows that the *h*-index of EMAG series since 1989 is 8, with an average citation per item of about 0.5 for the older editions (due to the larger number of papers published) and about 1.6 for the 2009-2015 editions. This suggests that the most recent Proceedings are proving relevant and useful, and worth publishing as a collection.

I would like to thank all the reviewers who have helped during the meeting, as well as all the authors who have contributed to the 2017 edition of the EMAG Proceedings. I am sure you will enjoy reading these papers which contain some of the highlights of our excellent conference.

Caterina Ducati

IOP Conf. Series: Journal of Physics: Conf. Series **902** (2017) 011002 doi:10.1088/1742-6596/902/1/011002

Peer review statement

All papers published in this volume of *Journal of Physics: Conference Series* have been peer reviewed through processes administered by the proceedings Editors. Reviews were conducted by expert referees to the professional and scientific standards expected of a proceedings journal published by IOP Publishing.

IOP Conf. Series: Journal of Physics: Conf. Series **902** (2017) 012001 doi:10.1088/1742-6596/902/1/012001

Observation of thermally-induced magnetic relaxation in a magnetite grain using off-axis electron holography

Trevor P. Almeida[1]*, Adrian R. Muxworthy[2], András Kovács[3], Wyn Williams[4] and Rafal E. Dunin-Borkowski[3]

[1]School of Physics and Astronomy, University of Glasgow, Glasgow, G12 8QQ, UK.
[2]Department of Earth Science and Engineering, South Kensington Campus, Imperial College London, London, SW7 2AZ, UK.
[3]Ernst Ruska-Centre for Microscopy and Spectroscopy with Electrons (ER-C) and Peter Grünberg Institute, Forschungszentrum Jülich, D-52425 Jülich, Germany.
[4]School of GeoSciences, University of Edinburgh, Edinburgh, EH9 3JW, UK.

Abstract. A synthetic basalt comprising magnetic Fe_3O_4 grains (~ 50 nm to ~ 500 nm in diameter) is investigated using a range of complementary nano-characterisation techniques. Off-axis electron holography combined with *in situ* heating allowed for the visualisation of the thermally-induced magnetic relaxation of an Fe_3O_4 grain (~ 300 nm) from an irregular domain state into a vortex state at 550˚C, just below its Curie temperature, with the magnetic intensity of the vortex increasing on cooling.

1. Introduction

Palaeomagnetists study the magnetic signal recorded by magnetic minerals within rocks to understand a wide range of geological problems, e.g., plate tectonic movements and geomagnetic field variation. Small, magnetically-uniform grains, termed single domain (SD), are considered the strongest remanence carriers with good stability over geological time periods [1]. However, the magnetic signature in most rocks is usually dominated by larger magnetic grains (~ 0.1 to 10 μm) that display non-uniform magnetic structures, i.e., single or multiple vortices, that are commonly referred to as pseudo-SD (PSD) particles. Magnetite (Fe_3O_4) is usually considered the most important magnetic mineral due to its high abundance and spontaneous magnetization. In igneous rocks, the leading source of magnetic induction is thermal remanent magnetization (TRM), which is acquired in the direction of the ambient geomagnetic field as the grains cool below their Curie temperature (T_C (~ 580˚C for Fe_3O_4)). However, current understanding of the thermomagnetic behaviour of PSD remanence is restricted to bulk magnetic measurements and numerical models. Hence, our knowledge of PSD remanence and magnetic stability as a function of temperature is poor, and understanding of reliability of most Earth and planetary paleomagnetic signals is limited.

The transmission electron microscopy (TEM) technique of off-axis electron holography allows for the nano-scale imaging and quantitative measurements of magnetic induction within and around materials as a function of temperature [2]. Recent studies have shown that PSD vortex states in synthetic Fe_3O_4 powdered grains are generally stable as they approach the T_C and recover their signal on cooling [3-5]. However, few studies have shown visually the thermoremanent behaviour of PSD Fe_3O_4 grains dispersed within a matrix, which are more representative of rocks. In this context, a synthetic basalt containing nano-scale Fe_3O_4 grains is investigated using a range of nano-characterisation techniques, and off-axis electron holography is combined with the heating of PSD magnetite (Fe_3O_4) grains *in situ* within the TEM.

*Author to whom correspondence should be addressed: trevor.almeida@glasgow.ac.uk.

IOP Conf. Series: Journal of Physics: Conf. Series **902** (2017) 012001 doi:10.1088/1742-6596/902/1/012001

2. Experimental

The synthetic basalt was produced via a glass-ceramic method by heating laboratory grade powders in a vertical temperature-controlled tube furnace under an oxygen partial pressure at 1400°C, before quenching and reheating at 750°C, as described previously [6]. The sample was mounted in epoxy resin and polished for secondary electron microscopy (SEM) using an FEI Quanta 200 ESEM FEG, whilst their elemental constituents were identified by energy dispersive X-ray (EDX) analysis. Conventional bright-field (BF) imaging was performed using a FEI Tecnai TEM operated at 200 kV. For structural characterisation, part of the synthetic basalt was crushed into a fine powder and deposited onto silicon substrates for crystallographic identification using X-ray diffractometry (XRD) (PANalytical X'Pert Pro Diffractometer). Magnetic measurements were conducted at Imperial College London and the Institute for Rock Magnetism, University of Minnesota. The first-order reversal curve (FORC) and high temperature thermomagnetic curves were measured using a Princeton Measurements Vibrating Sample Magnetometer (VSM) fitted with a furnace; heating was performed in flowing He. Low-temperature hysteresis measurements were made from 15 to 295 K (in 10 K intervals) using a helium-cooled Princeton Measurements VSM.

For *in situ* heating TEM investigations, desirable regions were deposited and welded with platinum to contacts on a DENSsolutions® double-tilt MEMS-based heating chip in a focused ion beam SEM before being milled into thin sections. Off-axis electron holograms were acquired at 300 kV in Lorentz mode in a Titan 80-300 TEM (ER-C) [8] with a charge-coupled device camera and an electron biprism, operated at 90 V. The directions of magnetization in the Fe_3O_4 particle were reversed initially at 20°C by tilting the sample ± 75° and turning on the objective lens to apply a magnetic field of > 1.5 T. The objective lens was then turned off, and the sample tilted back to 0° for hologram acquisition under field-free conditions with the particle magnetized in opposite directions, and the mean inner potential (MIP) was then separated from the magnetic potential [2-5]. Electron holograms were then acquired under field-free conditions (residual field < 0.2 mT) during *in situ* heating at 400°C, 500°C and 550°C, and again upon cooling. The heating experiment was then repeated, and magnetization reversal was performed by turning on the objective lens at ± 75° at each temperature interval to obtain the MIP, which was then subtracted from the total phase shift attained during the first heating stage to isolate the Fe_3O_4 particle's true remanent magnetisation [4, 5].

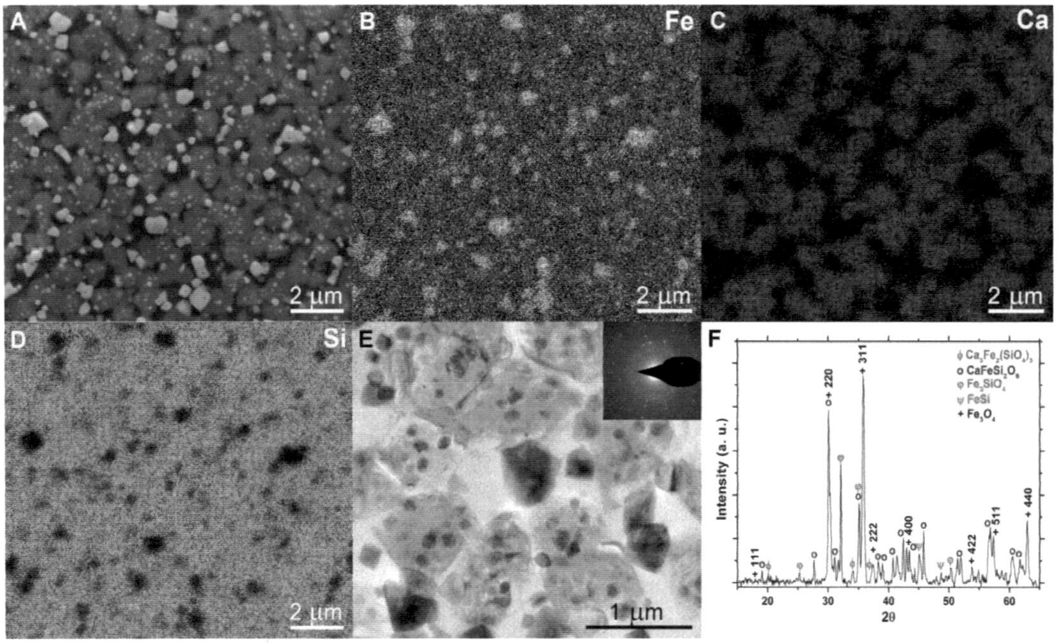

Figure 1. (A) BSE SEM image of the synthetic basalt; and (B-D) associated EDX chemical maps showing the elemental distribution of (B) iron; (C) calcium; and (D) silicon. (E) BF TEM image of the Fe3O4 grains within the silicate matrix ranging from ~ 50 to ~ 500 nm, with SAED (inset). (F) XRD pattern of the synthetic basalt, confirming the Fe_3O_4 phase (indexed, JCPDS 75-449), as well as the additional phases of $CaFeSi_2O_6$, $Ca_3Fe_2(SiO_4)_6$ and FeSi.

3. Results

The furnace annealing of the precursor powder mixture led to the production of a synthetic basalt comprising Fe_3O_4 grains of varied size, morphology and magnetic properties, the evidence for which is now presented in detail.

Fig. 1 provides information on the chemical distribution, morphology, size and structure of the synthesised sample. The backscattered-electron (BSE) SEM image and corresponding EDX maps of Fig. 1A-D present the elemental distribution within the sample. Comparison of the BSE image (Fig. 1A) with the Fe chemical map (Fig. 1B) confirm the bright regions in the BSE image are Fe-rich, and hence attributed as the Fe_3O_4 particles. Some areas are observed to be high in Ca content (Fig. 2C), whilst the remaining area is shown to be Si-rich (Fig. 1D). The BF TEM image of Fig. 1E displays Fe_3O_4 grains ranging in size, from ~ 50 nm to ~ 500 nm, along with the surrounding glass ceramic matrix. It is evident that the smaller Fe_3O_4 grains are generally confined within glassy regions, separate from the larger Fe_3O_4 grains. The peaks in the XRD pattern of Fig. 1F are in good agreement with the presence of Fe_3O_4 (Joint Committee on Powder Diffraction Standards (JCPDS) ref: 75-449), as well as pyroxene hedenbergite ($CaFeSi_2O_6$, JCPDS 70-1876), andradite ($Ca_3Fe_2(SiO4)_3$, JCPDS 84-1938), FeSi (JCPDS 86-0795) and Fayalite (Fe_2SiO_4, JCPDS 71-1673).

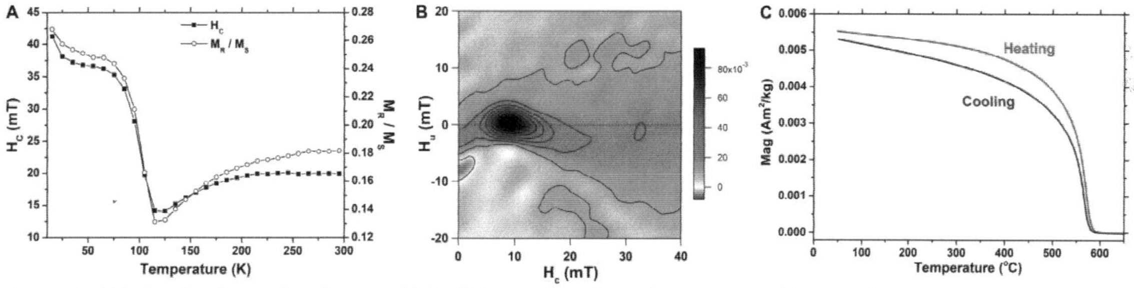

Figure 2. (A) Graph of coercive force and M_{rs} / M_s as a function of temperature (15 K to 295 K). (B) FORC diagram acquired at room temperature (smoothing factor = 7). The measurement time was 250 ms. (C) Thermomagnetic curves of the synthetic basalt. Heating was performed in flowing helium, in a field of 300 mT.

The hysteresis parameters H_c and M_{rs} / M_s are plotted as a function of temperature (15 K to 295 K) in Fig. 2A and displays behaviour typical of Fe_3O_4, with the Verwey transition at ~ 120 K. The FORC diagram of Fig. 2B shows the sample to exhibit behaviour characteristic of a mixture of SD and PSD grains, whereby the negative region close to the vertical H_u-axis indicates that the controlling anisotropy in SD grains is uniaxial, whilst the curvature along the H_c-axis reveals the presence of PSD grains. High-temperature thermomagnetic analysis (Fig. 2C) indicates that the Fe_3O_4 NPs within the synthetic basalt are thermally stable. The T_C was determined from the heating curve using the second-derivative method [8] and calculated as 585 ± 5°C, which is in good agreement with stoichiometric Fe_3O_4.

Fig. 3 presents the thermomagnetic behaviour of an individual Fe_3O_4 grain embedded in the silicate matrix as it is heated from 20°C to 550°C and then cooled back to 20°C. The BF TEM image of Fig. 3A displays the square Fe_3O_4 grain with sides ~ 300 nm in length and a hole located in its centre, with the associated selected area electron diffraction (SAED) pattern (inset). Fig. 3B displays a magnetic induction map of this Fe_3O_4 grain at 20°C after saturation, and reveals its magnetization to resemble a horseshoe-like shape, with the magnetic contours flowing in a clockwise direction, along with a component of stray magnetic field, indicative of a PSD state. An increase in temperature to 400°C (Fig. 3C) results in a slight widening of the contours and further broadening is evident in Fig. 3D, recorded at 500°C. At 550°C, the magnetization closes in to a vortex structure (Fig. 3E), the intensity of which becomes more prominent upon cooling to 500°C (Fig. 3F). The magnetic contours of the vortex are observed to narrow and the vortex magnetization intensity to increase further upon cooling back to 20°C (Fig. 3G & 3H). As a consequence of the vortex closure, it is evident that the stray magnetic field (Fig. 3H) has become notably less pronounced than observed prior to heating (Fig. 3B), i.e., it has partially demagnetised.

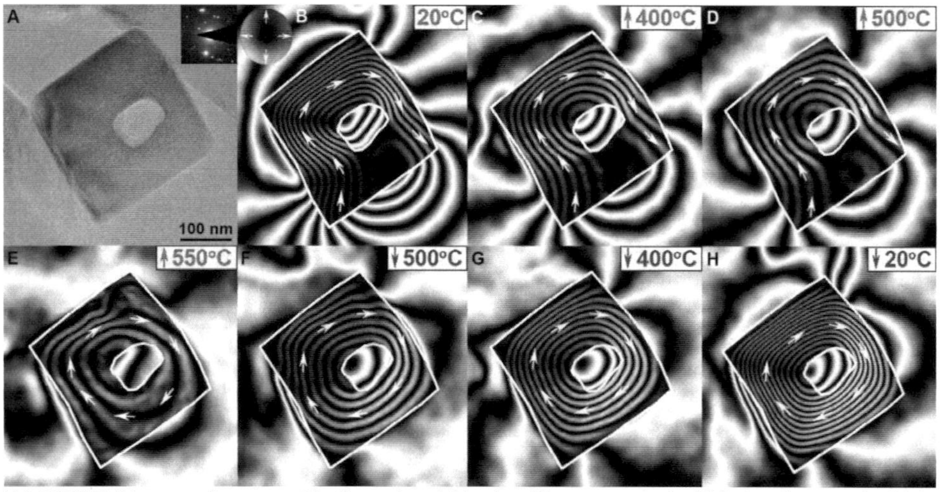

Figure 3. (A) BF TEM image of a square Fe_3O_4 grain within the silicate matrix (~ 300 nm along each side), exhibiting a hole at its centre, with associated SAED pattern (inset). (B-H) Magnetic induction maps reconstructed from holograms taken at (B) room temperature (20˚C); and during *in situ* heating to (C) 400˚C; (D) 500˚C; and (E) 550˚C; as well as upon subsequent cooling to (F) 500˚C; (G) 400˚C; and (H) 20˚C. The contour spacing is 0.39 radians for all magnetic induction maps and direction is shown using arrows, as depicted in the colour wheel.

4. Discussion

This combined SEM, TEM, XRD and magnetic investigation of the synthetic basalt has provided evidence for the growth of Fe_3O_4 grains, ranging from ~ 50 to ~ 500 nm in size, within the silicate matrix. The smaller 50 nm grains are confined within Ca-rich glassy regions, and hence considered as hedenbergite regions; these SD domain grains are considered to form during the rapid quenching stage of synthesis [6]. The larger grains, ~ 500 nm in diameter, are likely to contribute the PSD signal in the FORC diagram. Complementary to the presence of the Verwey transition in the Fig 2A, the thermomagnetic curve displays a T_C characteristic of stoichiometric Fe_3O_4, and shows a significant drop in magnetisation above 400˚C, providing a suitable temperature range for investigating the magnetic relaxation of PSD Fe_3O_4 grains using electron holography.

The Fe_3O_4 grain in Fig. 3 exhibits a horseshoe-like magnetisation state (Fig. 3B) that reduces in intensity upon heating to 400˚C and 500˚C, closing into a vortex structure at 550˚C, which then steadily increases in intensity upon cooling, with its stray magnetic field intensity reducing significantly compared to its initial state. It is apparent that the thermal energy imparted on the Fe_3O_4 grain at 550˚C has promoted the transition to a more energetically-favourable vortex state. For bulk samples, the measurable net directional and intensity information is acquired from the stray magnetic fields of the grains and it is evident this signal is reduced after vortex closure, losing both its original directional and intensity information. Hence, whilst the domain structure remains relatively stable during heating, at close to the T_C (> 500˚C) the original palaeomagnetic signal can be lost due to thermal relaxation.

In summary, complementary nano-characterisation techniques have confirmed the synthesis of SD and PSD Fe_3O_4 grains within a synthetic basalt. Electron holography combined with *in situ* heating allowed for the visualisation of the magnetic relaxation of an PSD Fe_3O_4 grain into a vortex structure just below the T_C.

References

[1] Néel L 1995 *Adv. Phys.* **4**, 191-242 L.
[2] Dunin-Borkowski R E *et al* 1998 *Science* **282**, 1868-1870.
[3] Almeida T P *et al* 2014 *Geophys. Res. Lett.*, **41**, 7041–7047.
[4] Almeida T P *et al* 2016 *Sci. Adv.*, **2**, e1501801.
[5] Almeida T P *et al* 2016 *Geophys. Res.* Lett., **43**, 8426-8434.
[6] Almeida T P *et al* 2014 *Geochem. Geophys. Geosyst.*, **15**, 161–175.
[7] Boothroyd C, Kovács A and Tillmann K 2016 *JLSRF* **2,** A44.
[8] Tauxe L 1998 Paleomagnetic Principles and Practice, (Kluwer Acad., Dordrecht, Netherlands).

Acknowledgements
The authors would like to thank NERC (grant NE/J020966/1) and ERC (grant 320832) for funding.

In-situ micro bend testing of SiC and the effects of Ga^+ ion damage

S Robertson, SS Doak, Z Zhou and H Wu

Department of Materials, Loughborough University, Leicestershire, LE11 3TU, UK.

s.robertson@Lboro.ac.uk

Abstract. The Young's modulus of 6H single crystal silicon carbide (SiC) was tested with micro cantilevers that had a range of cross-sectional dimensions with surfaces cleaned under different accelerating voltages of Ga^+ beam. A clear size effect is seen with Young's modulus decreasing as the cross-sectional area reduces. One of the possible reasons for such size effect is the Ga^+ induced damage on all surfaces of the cantilever. Transmission electron microscopy (TEM) was used to analyse the degree of damage, and the measurements of damage is compared to predictions by SRIM irradiation simulation.

1. Introduction

Micro mechanical testing has been widely enployed for the characterisation of materials used for micro- and nano- electro-mechanical systems (MEMS/NEMS). Young's modulus size effect is of interest because it is an essential input for finite element method (FEM) modelling [1]. Focused ion beam (FIB) has the ability to machine a range of geometries for testing bulk mechanical properties along specific crystallographic orientations, or properties of interesting regions such as grain boundaries [2], [3]. However, many micro mechanical tests have failed to match theoretically predicted properties, such as Young's modulus and breaking strength, despite their size being sufficiently small to contain few or no dislocations or flaws [4], [5]. A likely factor is the changes introduced by ion milling. Such changes may include the creation of an amorphous layer causing discrepancies in the measurement of the geometry of micro mechanical samples and the introduction of surface dislocations, point defects, and/or residual stresses [6]. SiC has great potential for NEMS/MEMS, especially for uses at high temperatures or in harsh environments. Single crystal SiC typically deforms elastically in bending until failure, is electrically conductive, and can be milled well in the FIB, making SiC an ideal model material for exploring the size effect and surface damage by a Ga^+ ions.

2. Methods / materials

The micro cantilever beams were prepared using a FIB system (Nova 600 Nanolabs, FEI, USA) in a single crystal of 6H SiC (Marketech International, USA). To ensure the top face of the beams were orientated parallel to the 0001 plane a three stage milling technique was developed. Once milled to the required dimensions all faces were finished by a Ga^+ beam under three different accelerating voltages and currents: 30 keV and 0.3 nA followed by 10 keV and 50pA for 20s; 30 keV and 0.3 nA followed by 5keV and 70 pA for 30s. Three beams were prepared for each finishing conditions. Bending test was completed inside the FIB through loading the micro beam with a micro probe (FMT 120, Kleindiek

Nanotechnik, Germany). The Y stage of the microscope was indexed towards the indenter bending the cantilever. A video and time stamped images were recorded to link force and displacement.

To gain evidence of the surface modification by the Ga^+, a cross sectional lift-out foil was prepared for TEM analysis from a surface finished with Ga^+ beam under each of the three aforementioned finishing conditions. Microstructure characterisation of the cross-section was performed using a FEI Tecnai F20 G2 S-Twin field emission gun (FEG) scanning TEM (STEM). Selected area electron diffraction patterns and conventional high-resolution electron micrographs (HREM) of the SiC were recorded. Energy dispersive X-ray (EDX) spectroscopy line scans were conducted using an Oxford Instruments X-Max 80mm2 TLE detector.

The Stopping and Range of Ions in Solids (SRIM, 2008, USA) calculations were completed with full cascade simulations up to 250 ions [7]. SiC was modelled with a density of 3.16g/cm^3 and a compound coefficient of 1, Gallium was model with a mass of 68.93 amu and an accelerating voltages of 30 keV, 10 keV and 5 keV [7]. Modelling provide ion range, straggle, sputter yield and vacancies per ion at a range of incident angles.

3. Results and discussion

Figure 1 shows a typical bend test in progress. From these still images the displacement and contact point of the beam were measured.

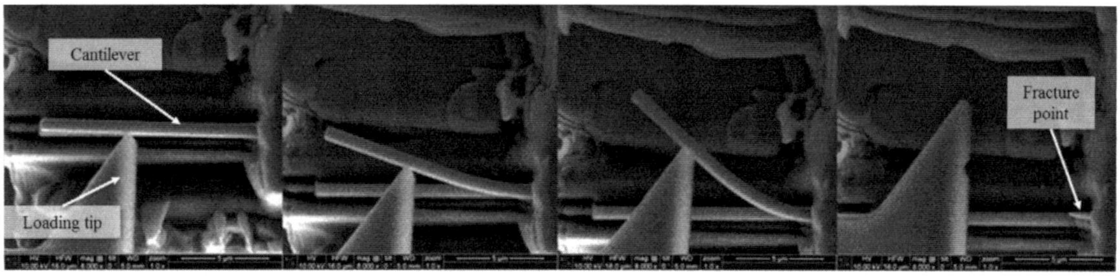

Figure 1. micro bend testing in progress left to right showing deformation to failure

From the force displacement data, small deflection beam equations were used to determine modulus and failure strength. Figure 2(a) summarizes the Young's modulus measurements obtained from SiC micro-cantilevers that have different cross-sectional areas and have undergone different surface cleaning conditions.

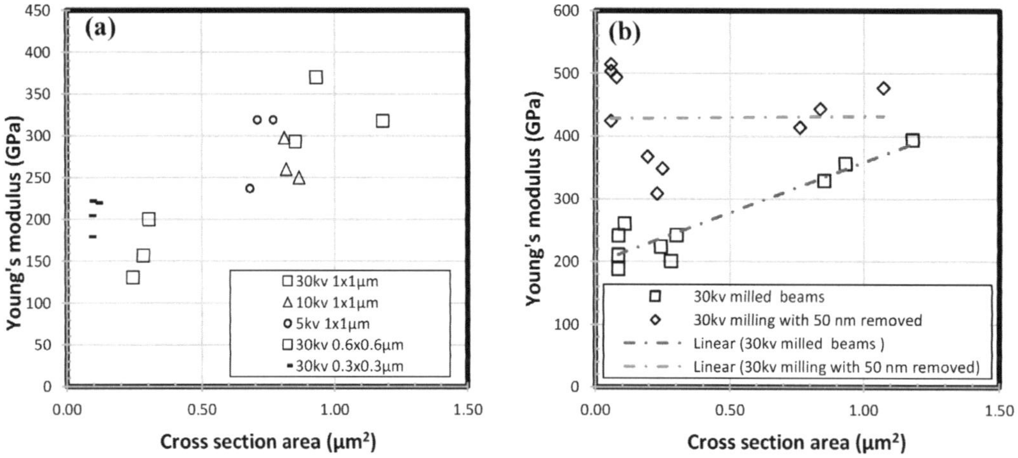

Figure 2. (a) micro bend test results modulus against cross section (b) adjusted micro bend tests data with original 30 kV data and beams with 50 nm removed from breadth and width

Young's modulus decreases approximately linearly as the cross-section area of a cantilever beam becomes smaller. The bulk modulus of SiC is cited as > 415 GPa [8]. Among our measurements, the highest Young's modulus, 393.98 GPa, was measured from beams with a cross-sectional of 1.18 μm^2, and the lowest, 187.84 GPa, from those with a cross sectional area of 0.086 μm^2. Lower accelerating voltage cleaning did not seem to have a strong impact on the size effect, as shown in Figure 2(a).

TEM analysis shows that the thickness of damaged layer by each cleaning routine was similar, 42-51 nm. The gallium ion beam produced an amorphous layer on the surface with a sharp transition between the damage layer and the bulk material, as shown in Figure 3. SRIM simulations predicted an ion penetration range of 19 nm at 30 keV (as shown in Figure 3(b)), nearly 3 times smaller than measured in the sample. EDX in STEM mode, shows the damaged layer contained around 1.5 At % Gallium at 21 nm depth from the surface, this matches well with the SRIM simulations.

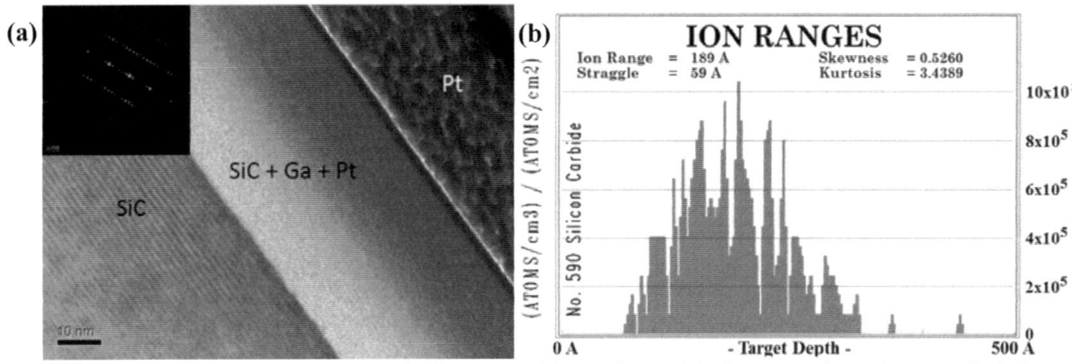

Figure 3 (a) shows HRTEM image of silicon carbide surface with the amorphous layer and platinum protective coating. (b) SRIM simulation of 30 kV gallium ion implantation distribution in silicon carbide

The mechanical performance of the damaged level is still under investigation, we believe that this layer likely contributes to the observed Young's modulus size effect. Other factors may contribute the size effect. Based on the TEM analysis, 50 nm was removed from the breadth and thickness of the micro beams and the modulus calculations were reprocessed to simulate the effects of the amorphous layer on the results of the micro bend test results. The effects of reprocessing the measurements from cantilevers with surfaces finished under 30 keV can be seen in Figure 2(b) where the modulus of the beams is increased closer to literature values. It was noted that cantilevers with a cross section of 0.3x0.3 μm exhibited plastic deformation at high loads. These beams give overestimated Young's modulus of the SiC, and the likely reason is due to the inability of the small deflection equations dealing with plastic deformation.

4. Conclusions

It was identified that a size effect is present in SiC when testing micro beams fabricated in the FIB. The source of the size effect may be from the amorphous zone developed on the surface of the material. Cleaning with low accelerating voltages did not significantly improve mechanical properties of the micro cantilevers. Based on SRIM calculations it is fair to assume that prolonged low voltage cleaning might have likely reduced the ion damage layer, however this strategy would also lead to the modification of the micro beams shape. Plasticity was observed in micro beams with cross sections below 0.3x0.3 μm, and the modulus of these micro beams was over estimated if the small elastic deflection equations are used. SRIM simulations did not accurately predict the extent of the damage in the SiC surface but were accurate in predicting the implantation depth of the gallium ions.

IOP Conf. Series: Journal of Physics: Conf. Series **902** (2017) 012002 doi:10.1088/1742-6596/902/1/012002

References

[1] M. F. Pantano, H. D. Espinosa, and L. Pagnotta, "Mechanical characterization of materials at small length scales," J. Mech. Sci. Technol., 2012.
[2] D. E. J. Armstrong, A. J. Wilkinson, and S. G. Roberts, "Measuring anisotropy in Young's modulus of copper using microcantilever testing," J. Mater. Res., 2009.
[3] A. Stratulat, D. E. J. Armstrong, and S. G. Roberts, "Micro-mechanical measurement of fracture behaviour of individual grain boundaries in Ni alloy 600 exposed to a pressurized water reactor environment," Corros. Sci., 2016.
[4] H. Bei, S. Shim, M. K. Miller, G. M. Pharr, and E. P. George, "Effects of focused ion beam milling on the nanomechanical behaviour of a molybdenum-alloy single crystal," Appl. Phys. Lett., 2007.
[5] H. Bei, S. Shim, E. P. George, M. K. Miller, E. G. Herbert, and G. M. Pharr, "Compressive strengths of molybdenum alloy micro-pillars prepared using a new technique," Scr. Mater.2007.
[6] A. Tseng, "Recent developments in nanofabrication using focused ion beams," Small, 2005.
[7] B. J. P. and L. U. Ziegler J F, "The Stopping and Range of Ions in Solids (New York: Pergamon)." 1985.
[8] S. Kiani, K. W. K. Leung, V. Radmilovic, A. M. Minor, and J. Yang, "Dislocation glide-controlled room-temperature plasticity in 6H-SiC single crystals," Acta Mater., 2014

IOP Conf. Series: Journal of Physics: Conf. Series **902** (2017) 012003 doi:10.1088/1742-6596/902/1/012003

In-situ irradiation of cerium precursors in TEM to study nanocrystal formation

Muhammad Sajid Ali Asghar[1*], Mohammed Mohammed Sabri[1,2], Zijian Tian[1], and Günter Möbus[1*]

[1]Dept of Materials Science and Engineering, University of Sheffield, Sheffield, UK
[2]Present Address: Department of Physics, Faculty of Science and Health, Koya University, University Park, Kurdistan Region, Iraq
m.sajid@sheffield.ac.uk & g.moebus@sheffield.ac.uk

Abstract Three of the most commonly used precursor chemicals for wet-chemical nano-ceria synthesis are examined by means of direct dry electron irradiation in TEM. Transformation reactions of micron-size carbonate, chloride, and nitrate grains into nanocrystallites (internal or external) are recorded in situ. Progress of possible redox-changes of cerium is tracked by EELS. We find a straight local oxidation reaction for carbonates, but external nanorod formation by condensation in the case of chlorides, while nitrates show a multi-stage complex redox behaviour.

1. Introduction
Cerium-oxide, i.e. ceria, is one of the most important rare earth metal oxides, due to its chemical properties, e.g. absorption and release of oxygen and change in oxidation state (Ce^{3+} / Ce^{4+}). Ceria nanoparticle (CNP) synthesis has been extensively studied via ex-situ chemical methods like low temperature (e.g. sol-gel) chemical precipitation, or hydrothermal/ high temperature chemical reaction, etc. Furthermore, ceria NPs have various applications in the fields of energy, catalysis, biomedical science, automotive, and environmental science. The most common laboratory process for ceria NP fabrication includes wet-chemical precipitation synthesis from dissolved precursors, such as cerium nitrate, cerium chloride or cerium carbonate. This process, triggered e.g. by NaOH, is flexible and can be tuned towards nanorod or cube formation and often includes metastable intermediate morphologies or phases. It is therefore interesting to examine dry transformations of the same precursor chemicals triggered by electron irradiation instead of forced hydrolysis, as this allows live in-situ observation. This latter technique continues on earlier studies of metal nanoparticle fabrication by irradiating Me-halides [1-3], or on the Ca-carbonate to lime transformation in TEM [4].

2. Materials and Methods
We have used three hydrated precursors of cerium, including cerium carbonate hydrate (CCH), cerium chloride heptahydrate (CCl), and cerium nitrate hexahydrate (CNH); all from Sigma Aldrich, UK. For in-situ electron irradiation we have used TEMs JEM 3010 with a thermionic LaB6 electron gun and JEM 2010-F with field emission gun respectively. The JEM 3010 microscope allows the irradiation of materials with higher total current and higher accelerating voltage at 300 keV. The JEM2010-F, allows for finely focused electron irradiation as well as chemical analysis via energy dispersive x-ray spectroscopy (Oxford Instruments EDXS), and electron energy loss spectroscopy (EELS, Gatan GIF).We have used manual mortar and pestle to grind the raw powders. All micron-sized fragments

are initially TEM-homogeneous, all appear amorphous and all three materials should initially share a Ce^{III} oxidation state. Support films of holy carbon are used except for carbonates where silicon nitride thin films are preferred.

3. Results and Discussion

The results of dry irradiation of the three examined raw precursor powders of cerium chloride / cerium carbonate / cerium nitrate show different behaviour. For instance the electron irradiation of Ce carbonate hydrate is most predictable and reproducible at both voltages: The original phase develops internal sub-5nm nano-crystals by nucleation and growth (Figure 1). The entire impact region (beam imprint) shows clearly as a speckled and darker (higher average Z) zone. EELS tracking confirms that Ce^{III} converts to Ce^{IV} and the HRTEM of the nucleated nanocrystals is compatible with crystalline fluorite CeO_2, while the carbon peak rapidly shrinks (suspended on Si-nitride thin film).

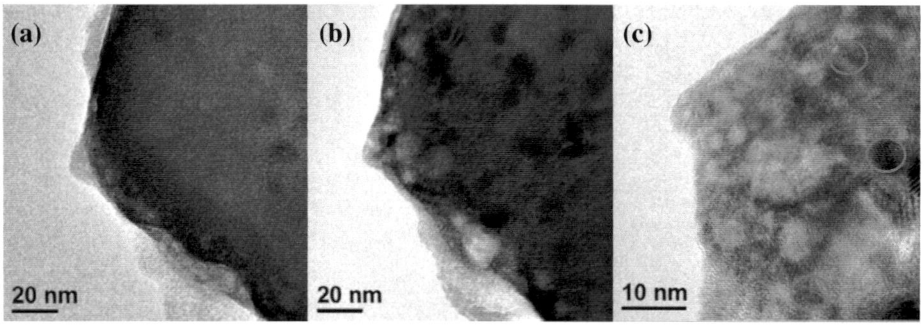

Figure 1: Electron irradiation of cerium carbonate: (a) before irradiation; (b) after 2min irradiation; (c) after 25min irradiation with crystallites encircled (JEM 2010F at 200kV).

The electron irradiation of cerium carbonate at 300 keV for two minutes shows a distinct impact region with formation of pores next to Bragg scattering new crystallites (Figure 2).

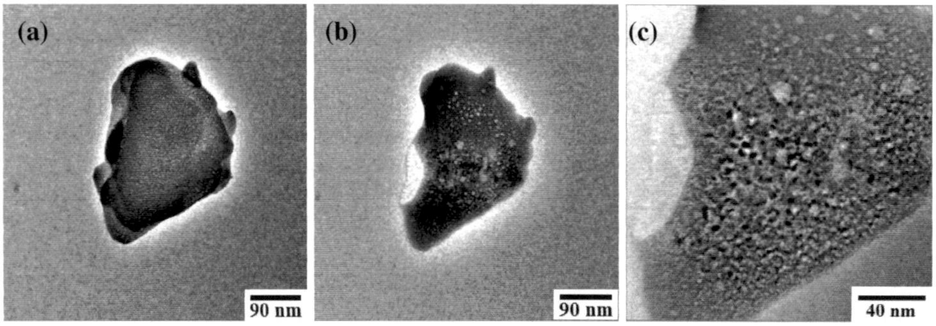

Figure 2: $CeCO_3$ particle (a) before and (b) after ~ 2 min of electron beam irradiation and (c) higher magnification image of the irradiated region from (b). JEM TEM 3010 at 300 keV.

Contrary, the cerium chloride hepta hydrate is found to respond much faster to irradiation but also more delocalised. Rather than internally precipitating, a vast number of fresh free-standing nanostructures are immediately formed on the adjacent carbon-support film or nanorods are growing out of the precursor grain as show in Figure 3. This is equivalent to an evaporation-condensation mechanism reported for Ag and metal-halides earlier [2, 3]. However, EELS confirms that Cl still persists in the new particles, which can be round or rod shaped, see Figure 3. Lattice fringe measurements on different particles match either $CeCl_3$ or CeO_2, however, EELS analysis is predominantly Ce^{III}.

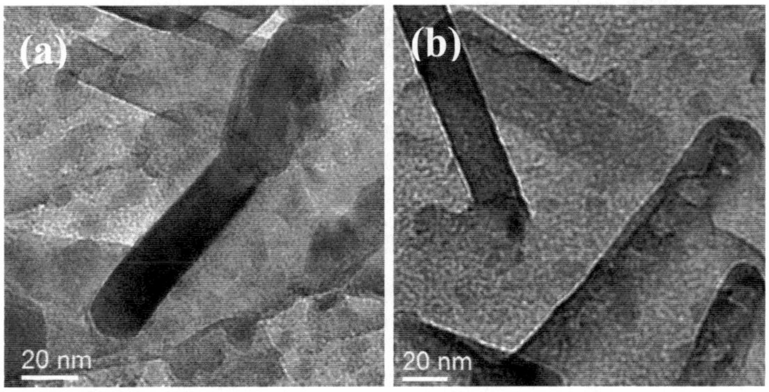

Figure 3: Two examples of irradiation of cerium chloride precursor: Nano-rods and some nanoparticles freshly formed on carbon film outside the precursor material (200kV)

Finally, the results of electron irradiation of cerium chloride at 300keV show the beam intensity to influence the shape of new condensation particles: moderate intensity leads to irregular shaped nanoparticles (Figure 4a-b) while high intensity leads to spherical nanoparticles (Figure 4d-e), possibly rounded trough ripening by secondary irradiation and quasi-melting.

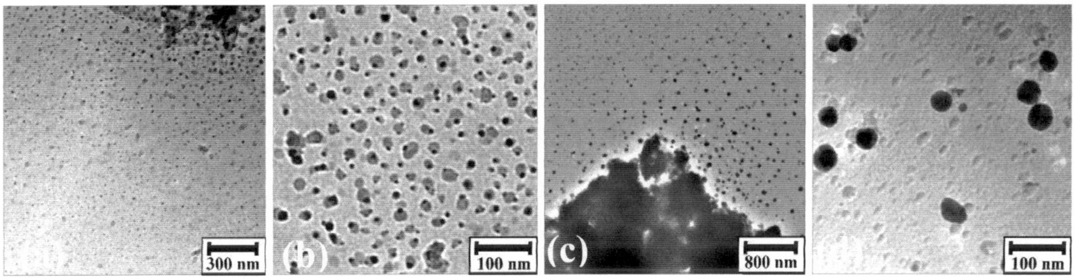

Figure 4: Irradiation of cerium chloride precursor: a) freshly formed irregular shape nanoparticles, c) roundish nano particles near a precursor fragment (higher intensity than (a)); (b, d) are magnified views of (a,c). All at 300kV.

Results of Cerium nitrate hexa hydrate superficially resemble the case of carbonate with similar appearance of beam impact zone and de-homogenisation. Figure 5 shows progress of transformation from amorphous state to nanocrystalline state, starting with a small presence of crystals, probably from imaging irradiation. However, EELS analysis failed to confirm simple nitrite to oxide conversion, as we find mainly Ce^{IV} converting to Ce^{III}, opposite to expectation.

We point to the great variety of redox behaviour found. Simple reductions of dry compound powders by electron beams are well established, such as for halide salts (e.g. AgCl to Ag [3]). Also constancy of valence has been found for the case of Ca-carbonate to lime [4]. Here we find a rare case of oxidation of cerium ions (unusual for TEM irradiation which normally involves reduction) for our selection of Ce-carbonate. In this case EELS results and HRTEM-spacings both match formation of fluorite CeO_2 phase as final stage.

Figure 5: Irradiation of Ce-nitrate precursor after deliberate focused irradiation with nano-crystallites forming (a) with FFT-pattern (b) matching CeO_2 phase.

For the Ce-chloride case, no redox reaction is confirmed, as the final crystalline particles and rods have all Ce, O, and Cl as elements, as evidenced by EELS, and Ce is in Ce^{III} state. This indicates that chloride rods and ceria particles might both form in parallel. Unlike any other precursor, the reaction progresses outside the original fragment, alike to the known delocal "evaporation/condensation" mechanism, known for other halides. However, no Ce-metal is found here. The unusual nitrate case with its redox-process appearing "inverted" from the carbonate case, might involve fast initial oxidation (before start of EELS monitoring) to CeO_2 followed by secondary beam-induced reduction to CeO_{2-x}, known to exist for this material, and for many other oxides [5].

4. Conclusion

Direct electron irradiation of dry precursor compound chemicals has a great potential to reveal insight into redox reactions, and is far too valuable to be labelled as unwanted "irradiation damage". Results are relevant for comparison with either laboratory-based wet-chemical redox change methods, but also with novel liquid-cell TEM based in situ methods (M S A Asghar et al., see elsewhere these proceedings).

References

[1] Malac M, Schoefield M, Zhu YM, Egerton R, J Appl Phys, (2002), 92, 1112-1121.
[2] Gnanavel T, Kumar S and Möbus G, J Nanosci Nanotechnology (2011), 11, 1019–1024.
[3] Li K, Zhang FS, J Nanopart Res, (2010), 12, 1423–1428.
[4] Golla-Schindler, U., et al., Microsc. Microanal. (2014), 20, 715-722.
[5] Long, N J and Petford-Long, A K, Mat. Res. Soc. Symp. Proc. (1987) **74**,299-304.

IOP Conf. Series: Journal of Physics: Conf. Series **902** (2017) 012004 doi:10.1088/1742-6596/902/1/012004

Ceria-Water-Reactions Studied by Liquid Cell TEM

Muhammad Sajid Ali Asghar, Beverley Inkson, Günter Möbus

Dept of Materials Science and Engineering, University of Sheffield, Sheffield, UK

m.sajid@sheffield.ac.uk, g.moebus@sheffield.ac.uk

Abstract. Liquid cell TEM is demonstrated to be a highly suitable new technology to study chemical reactions of cerium oxide nanostructures in aqueous environment. Ceria nanoparticles are found to show a diversity of behaviour. At low beam intensity, particle hopping movements and particle attachment, including chain formation is observable. At high intensity, when the water is turned ultra-enriched in ions by radiolysis, we find (in order of observation) particle rounding and shrinkage, followed by complete dissolution, along with formation of new reaction phases, and finally gas bubble formation.

1. Introduction

The unique advantage of in-situ TEM with a liquid-cell specimen holder is the capability of observing nanostructured materials in the original wet state, as they would mostly be employed in service. In fact, the wet-chemical aqueous synthesis is the most wide-spread and most high-throughput technique of all nanoparticle synthesis approaches. Furthermore, a majority of the most successful applications of oxide nanoparticles in consumer or industrial applications is in aqueous phase, including as examples, nanoceria polishing slurries, sunscreen lotions, combustion additives in transport fuels, wet-phase catalysis agents, environmental pollution cracking, and for chromium-replacement coatings on aluminium.

Standard dry TEM would have the disadvantage of missing intermediate phases in reaction studies (such as hydroxides unstable in air), but also cannot reveal the true aggregation behavior. Only liquid cell TEM of nanoparticle suspensions allows the observation of developments by continuous video recording. Often the radiolysis of water by intense electron beams is quoted as a limitation [1], as some reactions do not proceed as they would in neutral water. This work is however entirely based on using deliberate liquid cell irradiation of ceria suspensions as a study subject, and aims at emulating highly ion-enriched solutions and visualizing reactions under these extreme conditions.

2. Materials and Methods

Ceria nanoparticles are suspended in water and loaded into a liquid cell holder sandwiched between two Si-nitride membranes as described in [2] (model Protochips Poseidon P500 flow system used in static mode). TEMs chosen include JEM3010 LaB6 at 300kV and JEM2010F FEG with GIF at 200kV. Irradiation is then applied at variable levels with intensities adjusted to trigger the desirable reactions, see 3.1. Large changes to local water thickness means no universal dose rate threshold exists. Digital video recording allows for capture of sub-s frame rate, depending on binning, with mostly 6 frames/ s achieved in this work.

3. Results and discussion

Here we observe the behavior of ceria nano particles in water as a function of electron beam irradiation, which causes radiolysis and dissociation of molecule. For more details see [2, 3].

3.1. Corrosion/dissolution

Figure 1, row 1, illustrates the onset of dissolution of ceria nanoparticles, which can be tracked and quantified to reveal a monotonous volume loss. However, the dissolution rate (volume loss normalized to momentary surface area) is not constant and appears to accelerate [2]. Figure 1, rows 1 and 2, emphasizes the shape transformation from octahedral to spherical preceding the dissolution in many cases, confirming that corner atoms suffer faster dissolution rates than flat-faced atoms.

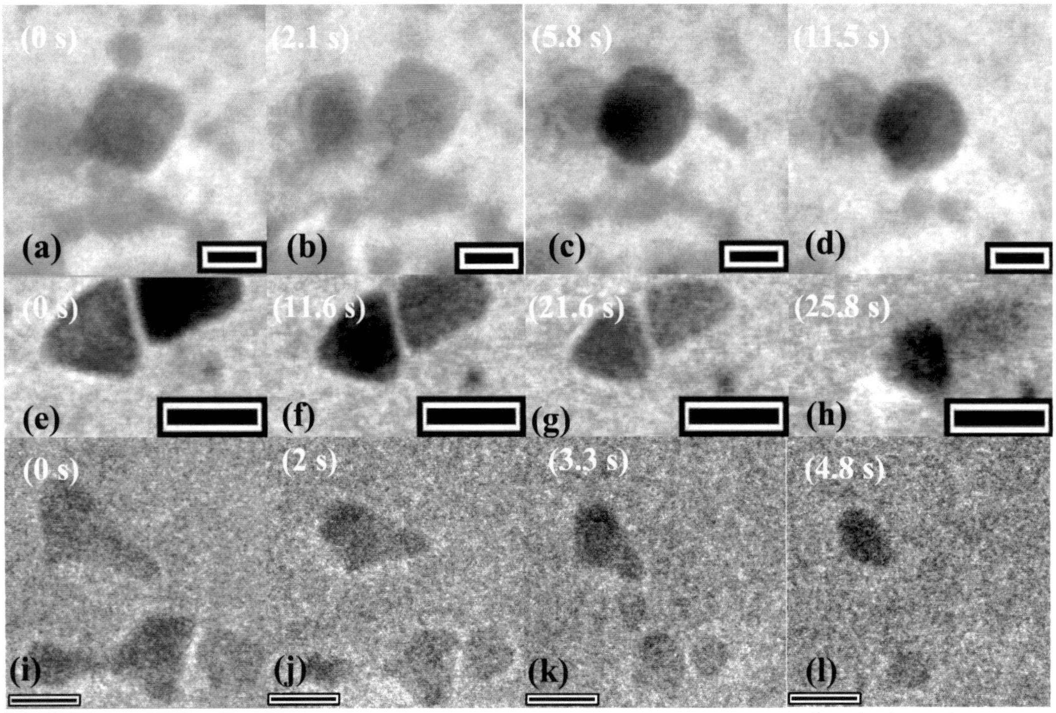

Figure 1. Octahedral ceria particles undergoing corner-ablation and rounding before dissolution (a-d and e-h); scale bar at 25nm. Final stages of dissolution of flat particles (i-l); scale bar at 10nm.

3.2. Chemical Reaction Phases

Figure 2 illustrates some of the new phases observed to appear after prolonged irradiation of the nanoceria-water suspension. The new phases are clearly secondary reaction products forming as a result of the water becoming saturated in dissolved ions from either surface hydrolysis reactions (cerium ions, OH- ions) or also from ions formed directly from irradiation induced water splitting without involvement of cerium oxide. In many cases, the order of formation is complex, and new phases grow while initial dissolution or shrinkage of ceria is still in progress or even in early stages. A common phase observed (Figure 2) appears to be rather flat, roundish and of low mass-thickness contrast / low average atomic number. Its structure never shows any Bragg-scattering features and is certainly amorphous. In analogy to ex-situ literature [4] we identify these particles as amorphous Ce-hydroxide. Figure 2, top row, shows a case in which the amorphous phase grows, while the last of the crystalline raw particles are still dissolving. Another interesting special case, Figure 2, bottom row, is the formation of bright bands around new particles (Figure 2e), possibly a Ce-ion depletion zone. Otherwise the new phase is rather stable against radiative dissolution.

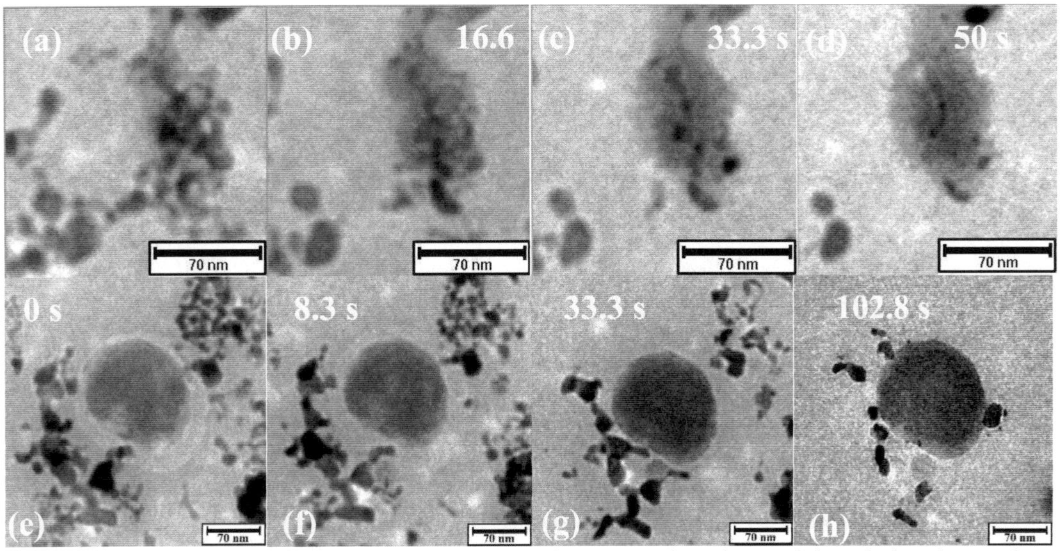

Figure 2. (a-d) New amorphous phase growing around ceria particles during later phases of dissolution. (e-h) amorphous phase initially surrounded by bright zone separating it from CNPs.

Further new phases observed [5] include the growth of crystalline high-aspect ratio needles during and alongside ceria dissolution, while another new phase after prolonged irradiation and dissolution appears roundish and of atomic number density similar to ceria. The final step in the sequence of experiments under deliberate high-intensity irradiation is the formation of a large number of nanosize bubbles. These gas bubbles of presumably [1] molecular hydrogen are of interest as in a ceria-water-suspension, two competing path-ways exist for H_2-production, (i) catalytic water splitting on ceria-surface active sites, and (ii) radiolytic direct water splitting anywhere in the liquid [6]. Our sequence of observations point to effect (ii) being dominant, while both might co-exist.

3.3. Exploring the 3D volume of liquid cell samples

More than any other kind of TEM specimens, liquid cell sandwich samples are intrinsically 3D, and a 3D reconstruction technique would be highly desirable. Standard tomography is not applicable, as tilt limits of the bulky holder, and more even self-shadowing of the deep Si-trough prevents high angular range series. Long series recording is also too slow, as particles might move rapidly.

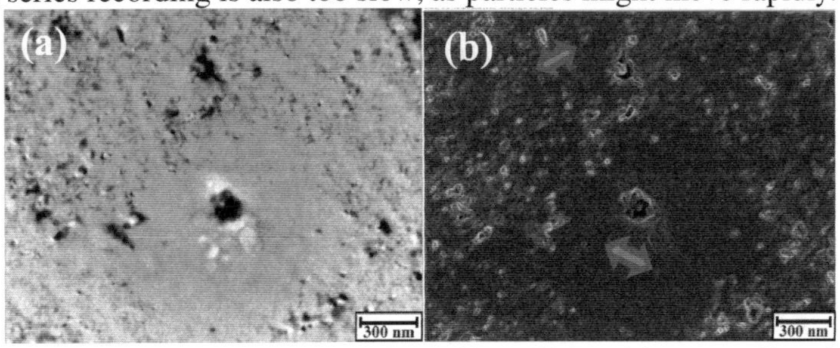

Figure 3. Raw image and superposition image from tilt-wobbler video recording showing two differing particle displacement directions (see arrows) and amplitudes.

(i) We propose in Figure 3 the acquisition of a video-sequence of nanoparticles suspended in water with the beam-tilt wobbler running at maximum speed. Particles attached to the top and bottom membrane clearly separate by their oscillation width and direction. Any Δz-height of floating particles can be extrapolated from their Δxy separation. An undisturbed video can still be recovered by selecting frames at one beam tilt only.

(ii) Another option is to evaluate Fresnel-fringe appearance around particles in vertical motion and calibrate the changes via a z-series of images manually lifted out of focus. (Figure 4)

(iii) While above methods (i,ii) result in the combined separation of the two membranes by water and gas bubbles, recording the EELS plasmon range signal from water [7] would give exclusive access to the water thickness only, as illustrated by a recording of spectra in different sample areas of increasing water thickness, showing Δz sensitivity through multiple plasmon scattering (Figure 5).

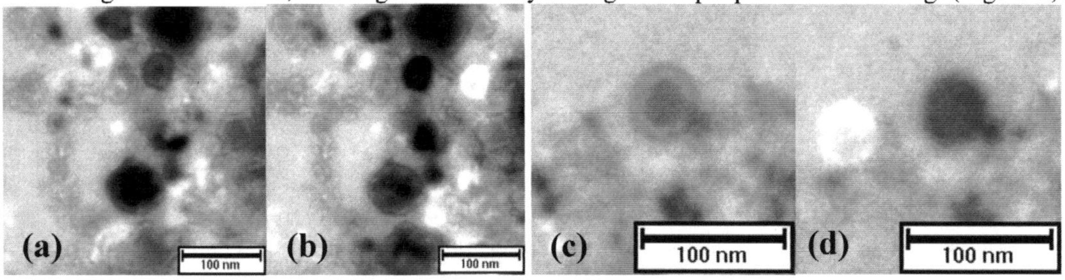

Figure 4. Pairs of particles showing jumping/tilting behaviour via Fresnel fringe changes.

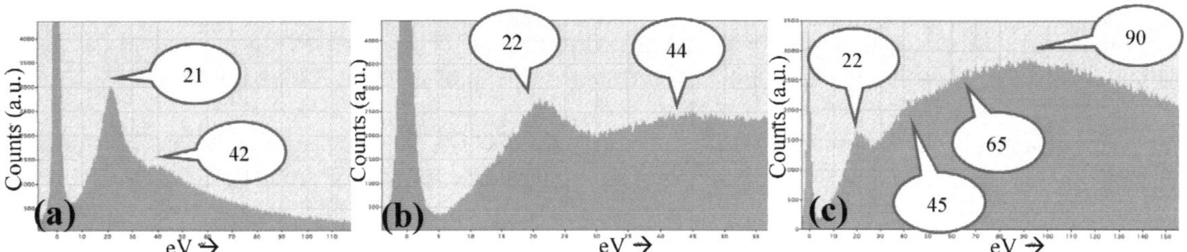

Figure 5. Low loss EELS of water at 3 thicknesses. Labels: single to 4-fold plasmon peaks (in eV).

4. Conclusions

The details of multiple reaction pathways, including dissolution of raw ceria nanoparticles and growth of secondary phases, with life video-recording of the nanoscale morphology development, are presented. The diversity of observations confirms that liquid-cell TEM in deliberately highly irradiated water should not be discarded as artificial, but be considered a highly useful testing methodology. Radiolysed water can be seen as an accelerated testing tool, allowing to research reactions which would run orders of magnitude slower (and being unobservable on the nanoscale) in laboratory wet-chemical setup. We also suggest that the quantification via time-resolved measurements of particle geometry data adds new precision to corrosion/reaction studies, which at the macroscale often suffer from invisible nanoscale surface roughness. Testing of multiple materials should also lead to faster classification of durability relative to one-another, as the rank-order would not shuffle, even if the radiolytic corrosion process is more complex than standard acid-leaching experiments.

The irradiation chemistry of aqueous ceria in particular gains importance due to the material being used or proposed as cell-protection agent during radiotherapy, irradiation-enhanced photochemical pollutant cracking in environmental waste-waters, and finally for CeO_2 as UO_2-simulant in nuclear materials testing, whether for nuclear fuels in service, or for long term immobilisation.

References

[1] Schneider, N.M. et al., 2014, J. Phys. Chem. C, **118**, 22373-22382
[2] Asghar, M.S.A. et al. 2017, ChemPhysChem, **18**, 1-6.
[3] Asghar, M.S.A. et al. submitted for publication
[4] Heisbourg, G. et al., 2004, *Journal of Nuclear Materials*, 335, 5–13.
[5] Williamson, M.J. et al., 2003, Nature Materials, **2**, 532-536
[6] Caër, S.L., 2011, Water, **3**, 235-253.
[7] Holtz, M.E., et al., 2013, Microsc. Microanal., **19**, 1027–1035.

IOP Conf. Series: Journal of Physics: Conf. Series **902** (2017) 012005 doi:10.1088/1742-6596/902/1/012005

A Quantitative Evaluation of Electron Beam Sensitivity in Calcite Nanoparticles

R W M Hooley [1], A P Brown [1], A N Kulak [2], F C Meldrum [2], R M D Brydson [1]
[1]School of Chemical and Process Engineering, University of Leeds, Leeds, LS2 9JT, UK
[2]School of Chemistry, University of Leeds, Leeds, LS2 9JT, UK
Email: pmrh@leeds.ac.uk

Abstract We present analysis and quantification of electron beam sensitivity in calcite in order to identify damage thresholds under which bright field TEM imaging, selected area electron diffraction and scanning TEM should be performed. A large reduction in damage under TEM was found when operating at 300 kV rather than 200 kV, suggesting that the irradiation induced degradation of calcite to calcium oxide is dominated by radiolysis. At 300 kV, bright field STEM imaging was able to retain lattice information at higher fluences than was possible using TEM and, although damage was still seen to occur, there was no observation of the formation of crystalline calcium oxide.

1. Introduction

Calcium carbonate or calcite is of widespread interest as a structural biomaterial and for its environmental and industrial importance, which has led to extensive research into its formation and properties. Understanding its structure and chemistry at the nanoscale is of interest especially in relation to its interaction with other components in complex mixtures, including fuel additives and pharmaceutical formulations. Whilst a heavily researched material, calcium carbonate is well known for its sensitivity under electron irradiation in both conventional and scanning TEM damaging through its radiolytic decomposition to calcium oxide. Although recent reports have provided partial quantification of the damage process [1-3], the purpose of this study is to use calcite as a model material to compare damage under different conditions and operating modes (e.g. TEM and STEM) and develop damage limitation strategies which may also be applicable to other electron beam sensitive systems [4,5].

2. Experimental

2.1. Materials

Calcite nanoparticles were produced through the carbonation of calcium oxide suspended in water [6]. Production of the calcite polymorph was confirmed by Raman spectroscopy and X-ray diffraction (data not shown). The particles were dried and then suspended in ethanol prior to deposition onto a either a holey carbon or lacy carbon film (Agar Scientific Ltd) for 200 kV experiments, and an ultrathin amorphous carbon or graphene oxide film (EM Resolutions Ltd) for 300 kV experiments, all supported on copper TEM grids.

2.2. 200kV TEM irradiation study

TEM irradiation experiments were carried out at 200 kV in a FEI Tecnai F-20 FEG-TEM with an Oxford Instruments X-Max 80 mm^2 EDX detector and a Gatan Orius CCD. Controlled irradiation at several electron fluxes from 2.5x10^4 e$^-$nm^{-2}s^{-1} to 1.6x10^5 e$^-$nm^{-2}s^{-1} was applied to the calcite particles, bright field (BF) images and selected area electron diffraction patterns (SAEDP) were taken at timed intervals between 5 and 15s. Prior to imaging and diffraction, the electron beam was blanked and the flux reduced to avoid unintended irradiation. The presence of polycrystalline calcium oxide in SAEDPs was investigated by focussing on the CaO (111) and (200) diffraction rings. SAEDPs were processed using the "reveal weak reflections" tool in the Difftools plugin for Gatan Digital Micrograph [7]. To investigate the loss of CO_2, a cluster of calcite particles was continually irradiated at a flux of 2.6x10^4 e$^-$nm^{-2}s^{-1} and successive EDX spectra (with an acquisition time of 30 s) were collected.

2.3. Comparing irradiation between TEM and STEM

A comparative study of electron beam sensitivity in STEM and TEM was performed in a FEI Titan[3] Themis G2 operated at 300 kV on calcite nanoparticles. BF-TEM images were taken with an electron flux of 3.8×10^5 $e^- nm^{-2} s^{-1}$ at 380×10^3 times magnification on a Gatan OneView CCD. BF-STEM images were taken with a probe current of 40 pA, 16 µs dwell time and a pixel size of 57.8 pm giving a fluence of 1.22×10^6 $e^- nm^{-2}$ per scan. The FFTs of both TEM and STEM micrographs were compared to investigate changes in crystallinity induced by electron irradiation.

3. Results and Discussion

3.1. 200 kV irradiation

BF-TEM imaging revealed production of what appear to be pores around the calcite particle edges, seen after accumulated fluences of approximately 1.9×10^6 $e^- nm^{-2}$, irrespective of electron flux (Figures 1B and 2A). Using SAED, the nucleation and growth of nano-crystalline CaO was found to manifest as polycrystalline rings in calcite diffraction patterns at fluences of 3.9×10^6 $e^- nm^{-2}$ (Figure 1B inset), also irrespective of electron flux (figure 2A). Dark field TEM imaging shows the presence of calcium oxide crystallites preferentially around the outer edges of the irradiated calcite particles (Figure 1C). The pores induced by irradiation may act as sites for the build-up and release of CO_2 from the particle bulk and as even as nucleation sites for calcium oxide crystallite formation.

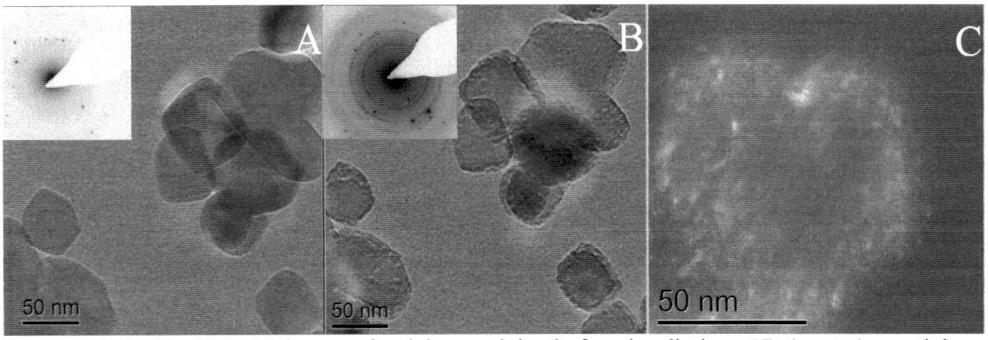

Figure 1A (left)- BFTEM image of calcite particles before irradiation, **1B** (centre)- particles after 1.9×10^7 $e^- nm^{-2}$. **1C** (right)- Dark field TEM image of particle after prolonged irradiation, showing calcium oxide nano-crystallites around the edge of the particle. Inset SAEDPs in Figure **1A** correspond to a cluster of particles before and in Figure **1B**, after an exposure of 1.1×10^7 $e^- nm^{-2}$.

A summary of the 200 kV electron fluences required for pore formation and CaO nucleation is shown in Figure 2A. Whilst these values do not exactly pinpoint the threshold fluence for the production of CaO, it does identify the point at which the material can be identified as being no longer purely calcite, thus diffraction information is compromised.

The decomposition of calcite has also been probed by EDX and this shows decay in the oxygen content (Figure 2B). This is consistent with the expected loss of CO_2 and, given that it would be expected to occur near the surface of particles, correlates with the DF-TEM image of crystalline CaO at the surfaces (Figure 1C). The rate of loss of CO_2 slows at higher fluences as alteration occurs in the particle cores leading to diffusion of CO_2 through the bulk to either the surface or a pore. The EDX data suggest that some mass loss is occurring before damage is detectable in BF imaging as pore formation and the appearance of polycrystalline CaO in SAED patterns are only identified at fluences of 1.9×10^6 $e^- nm^{-2}$ and 3.9×10^6 $e^- nm^{-2}$ respectively (Figure 2B). Thus we suggest that EDX measurements should be taken promptly, to avoid unrepresentative spectra.

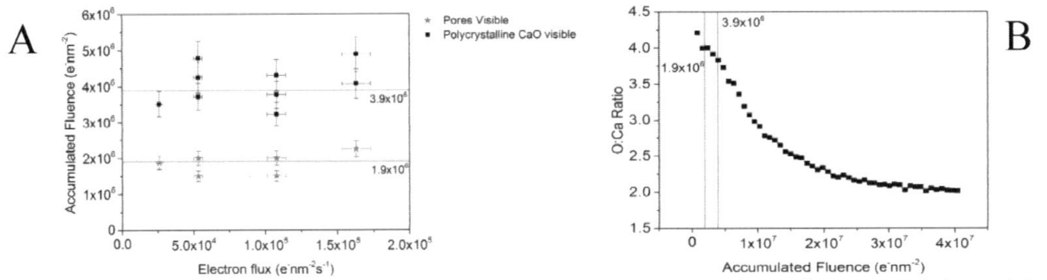

Figure 2A (left) – 200 kV electron fluence required for pores to appear in calcite nanoparticles and for the emergence of polycrystalline CaO rings in SAEDP for several electron fluxes. **2B** (right) – EDX derived atomic ratio of oxygen to calcium under continuous irradiation at $2.6x10^4$ $e^-nm^{-2}s^{-1}$, 200 kV. The initial low fluence O:Ca atom ratio is shifted from 3:1 to 4.2:1 presumably due to errors in the oxygen k factor used for EDX quantification. Using a correction factor to account for this implies that the O:Ca atom ratio plateaus at ~1.5 at high fluences i.e. full conversion to CaO is not achieved or CO_2 is still trapped in pores within the material.

3.2. 300 kV TEM and STEM irradiation

Under 300 kV TEM irradiation the FFT of the BF TEM image begins to show degradation of the calcite lattice, on average at a fluence of $5.5x10^7$ e^-nm^{-2} (Figure 3). This indicates that the extent of radiolysis is lower than at 200 kV, where the comparable damage threshold is $3.9x10^6$ e^-nm^{-2} (Figure 2A). This is an improvement of 14 times, despite the fact that knock-on damage may become a more significant factor at 300 kV.

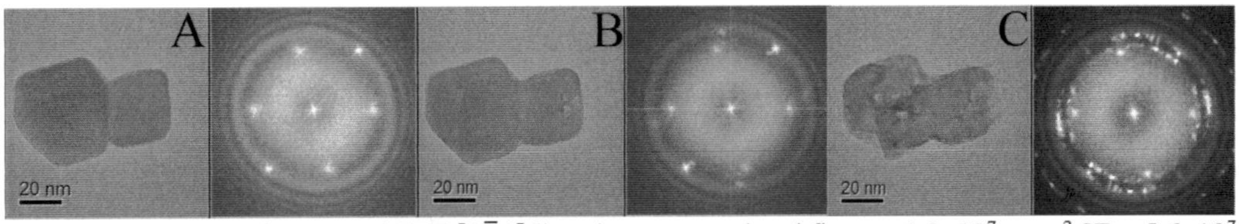

Figure 3A –BFTEM image and FFT $[4\bar{4}1]$direction, accumulated fluence $1.3x10^7$ e^-nm^{-2} **3B** – $5.6x10^7$ e^-nm^{-2} **3C** - $1.35x10^8$ e^-nm^{-2}.

Under STEM irradiation, crystalline CaO is not visible in the FFT of the BF-STEM image, even up to a fluence of $1.8x10^8$ e^-nm^{-2}, (Figure 4), suggesting that STEM may be a better choice when imaging calcium carbonate based materials. Features are seen in the STEM images throughout the irradiation series that are seen to move to the edges of the particle and disappear (white spots in Figure 4B), these may be gas bubbles trapped under the contamination layer that reach the edge of the particle and dissipate into the vacuum.

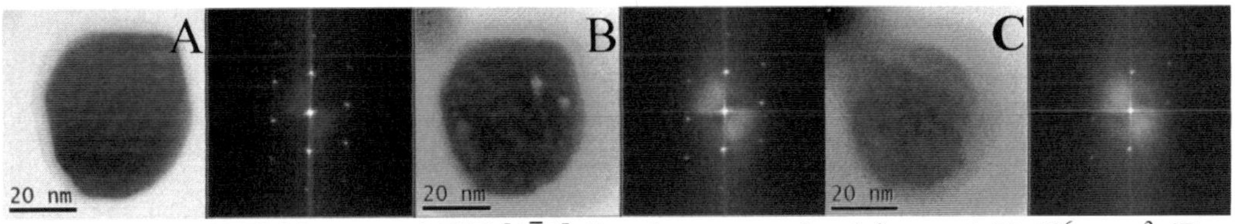

Figure 4A – BFSTEM image and FFT $[4\bar{4}1]$ direction, accumulated fluence $1.22x10^6$ e^-nm^{-2} **4B** – $5.5x10^7$ e^-nm^{-2} **4C** - $1.87x10^8$ e^-nm^{-2}.

The particle is still damaging under STEM irradiation, as the particle is seen to change shape over prolonged irradiation (Figure 4A and 4C) and bubbles appear under the contamination layer, however no crystalline CaO is observed, indicating either limited radiolysis, or the production of non-crystalline damage products. Hydrocarbon contamination should be considered as a contributing factor in the reduction of beam damage, as the contamination layer may be reducing the mobility of damage products, or reducing the incident electron flux.

The reduction of the calcite TEM FFT signal as calcite degrades with 300 kV irradiation matches well with the identification of CaO in SAEDPs, which is seen to increase in intensity after ~5.5×10^7 e$^-$ nm^{-2} (Figure 5A). Thus any phase contrast lattice imaging or selected area diffraction at 300 kV should be completed before this fluence is reached. Quantification of the reduction in spot intensity in the FFTs of BF-STEM images of calcite over prolonged irradiation does not show a threshold nor is the formation of CaO identified (Figure 5B). Thus alteration occurs either due to reduced crystallinity, and/or the build-up of hydrocarbon contamination. We note that the amorphous background, seen in the FFTs (Figure 4A-C) is seen to increase in intensity under irradiation.

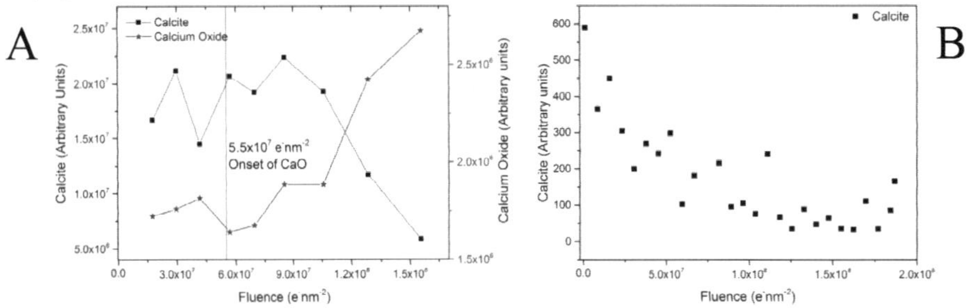

Figure 5A (left) - Integrated intensity of calcite $(01\bar{4})$ spot and section of CaO (111) ring under 300 kV TEM irradiation **5B** (right) - Integrated intensity of calcite $(01\bar{4})$ spot under 300 kV STEM.

4. Conclusions

Identification of electron fluence thresholds for different TEM analytical techniques can be prioritised based on the extent of material degradation that is deemed acceptable. For example, we have shown that in calcite, irradiation at 200 kV by bright field imaging should be undertaken at fluences below 1.9×10^6 e$^-$nm^{-2} and diffraction below 3.9×10^6 e$^-$nm^{-2}. EDX measurements should be taken with as little pre-irradiation as possible due to the rapid loss of CO_2. When operating at 300 kV rather than 200 kV, the TEM damage threshold of calcite increased significantly, from 3.9×10^6 e$^-$nm^{-2} to 5.5×10^7 e$^-$nm^{-2}, due to the reduction in radiolysis damage, caused by the reduction of the inelastic cross section. STEM of calcite at 300 kV however, was able to preserve crystallinity and lattice information up to higher fluences compared to both 200 kV and 300 kV TEM. There was no observable degradation into crystalline CaO at STEM fluences up to 1.87×10^8 e$^-$nm^{-2}. In order to fully compare electron beam damage in STEM and TEM, it is necessary to eliminate hydrocarbon contamination, to avoid any influence on the effects of electron beam irradiation.

These thresholds will be used during the structural and chemical analysis of calcium carbonate reverse micelles, currently used industrially as fuel additives [8].

Acknowledgements

RH acknowledges a studentship funded through the EPSRC CDT in Complex Particulate Products and Processes (cP^3-CDT) and acknowledges co-funding from Thermo-Fisher Scientific (formerly FEI Co).

References

[1] Hoffmann R. et al, *Micron*, 2014, **62**, 28-36.
[2] Golla-Schindler U. et al, *Microscopy and microanalysis* 2014, **20**, 715-722.
[3] Kovarik L. et al, *Applied Physics Letters*, 2016, **109**, 164102.
[4] Egerton R. F. *Ultramicroscopy*, 2013, **127**, 100-108.
[5] Egerton R. F. et al, *Micron*, 2004, **35**, 399-409.
[6] Hari B. et al, *Materials Letters*. 2006, **60**(12), pp.1515-1518.
[7] Mitchell D.R.G. *Microscopy research and technique, 2008* **71** 588-93.
[8] Mansot J. et al, *Colloids and Surfaces A*. 1993, **71** (123-134).

Cryo-STEM-EDX spectroscopy for the characterisation of nanoparticles in cell culture media

M Ilett[1], F Bamiduro[1], O Matar[1], A Brown[1], R Brydson[1] and N Hondow[1]

[1] School of Chemical and Process Engineering, University of Leeds, Leeds LS2 9JT, UK

pmmi@leeds.ac.uk

Abstract. We present a study of barium titanate nanoparticles dispersed in cell culture media. Scanning transmission electron microscopy combined with energy dispersive X-ray spectroscopy was undertaken on samples prepared using both conventional drop casting and also plunge freezing and examination under cryogenic conditions. This showed that drying artefacts occurred during conventional sample preparation, whereby some salt components of the cell culture media accumulated around the barium titanate nanoparticles; these were removed using the cryogenic route. Importantly, the formation of a calcium and phosphorus rich coating around the barium titanate nanoparticles was retained under cryo-conditions, highlighting that significant interactions do occur between nanomaterials and biological media.

1. Introduction

The promise shown by nanomaterials within medical research remains restricted by the difficulty in fully characterising nanoparticles within complex biological media [1]. Consequently, novel approaches are required to ensure nanomaterials can be fully characterised within this complex media. Commonly, *in vitro* investigations require nanomaterials to be dispersed in cell culture media for cell uptake studies. Therefore, the need to successfully monitor the nano-bio interactions that occur between nanoparticles and complete cell culture media (CCCM) has been identified.

Here we present an investigation looking at barium titanate (BT) nanoparticles dispersed in CCCM. BT nanoparticles have shown promise for use within medical imaging, cancer therapy, drug/gene delivery and tissue engineering [2]. Accordingly, a comprehensive understanding of how the nanoparticles interact with CCCM is required to better understand the results from *in vitro* cellular exposure experiments.

Analytical transmission electron microscopy (TEM) is one of the leading techniques for nanoparticle characterisation, with energy dispersive X-ray (EDX) spectroscopy used to provide elemental analysis. EDX spectroscopy can be combined with scanning TEM (STEM) in order to provide spatially resolved elemental maps. However, analytical TEM is often limited by the requirements during sample preparation for both thin and dry samples. Here we report on STEM-EDX spectroscopy carried out on BT in CCCM suspensions via both conventional drop casting (DC) TEM and plunge freezing with examination under cryogenic (cryo-) conditions. The results indicated a clear difference between the two techniques. This confirmed that drying artefacts can heavily influence the TEM analysis of nanomaterial suspensions, and highlighted analytical cryo-TEM as a viable option to combat this. In addition a significant interaction between the BT nanoparticles and CCCM was observed.

2. Experimental

2.1. Materials
BT nanoparticles, 130 nm in diameter were prepared by a previously reported hydrothermal method [3] and extensively characterised [4]. Dulbecco's Modified Eagles Medium (DMEM) (Sigma Aldrich, 51435C-500 ml Expiry date: Sep 2017) was supplemented with 1% penicillin-streptomycin (Sigma Aldrich, Lot# 066M4791V, P4333-100 ml) and 10 % Fetal bovine serum (FBS) (Sigma Aldrich, F7524-50 ml) and is referred to as CCCM throughout this paper.

2.2 Sample preparation
BT in CCCM stock suspensions were made at a concentration of 1 mg/ml by weighing the requisite amount of BT in 10 ml CCCM and then sonicating for a minimum of ~12 h. This stock was then diluted down to 100 μg/ml with further CCCM for analytical TEM analysis. For DC TEM grid preparation a holey carbon film TEM grid was plasma cleaned to make it hydrophilic and the BT-CCCM suspension drop cast onto it and allowed to air dry.

Cryo-TEM sample preparation was carried out using a FEI Vitrobot©. A Quantifoil TEM grid (R1.2/1.3; EM Resolutions) was plasma cleaned and then held within the Vitrobot© before a 3.5 μl drop was pipetted on the grid. The grid was then blotted and plunged into liquid ethane, after which it was maintained under liquid nitrogen before being transferred to a Gatan 914 TEM cryo-holder. The Vitrobot© settings were as follows; 100 % humidity, 30 s wait time and blot force 6.

2.3 TEM imaging and characterisation
For the drop cast specimens a FEI Tecnai F20 microscope operated at 200kV and fitted with a Gatan Orius CCD camera and Oxford instruments 80 mm^2 X-max EDX detector was used. For all cryo samples a FEI Titan^3 Themis G2 operating at 300 kV TEM fitted with 4 EDX silicon drift detectors, and a Gatan One-View CCD was used. For cryo-STEM-EDX imaging, conditions were set at dwell time 25 μs and probe current of 40 pA.

3. Results and Discussion
From TEM imaging of DC BT dispersed in CCCM a significant interaction at the nano-bio interface was identified, since a clear coating had formed around the nanoparticles (Figure 1(a)).

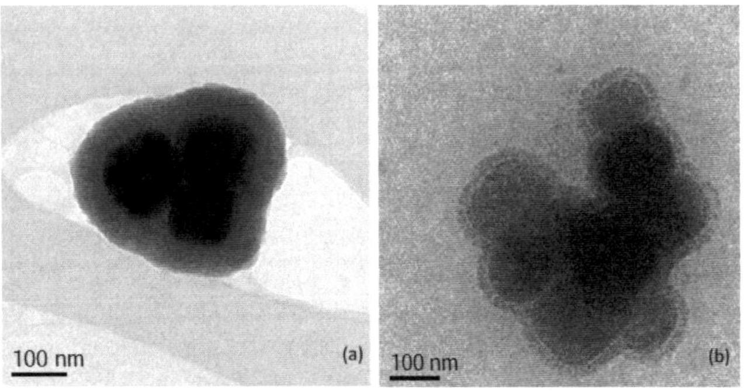

Figure 1: (a) TEM image of DC BT in CCCM showing a coating had formed around the nanoparticle agglomerates. (b) This was confirmed in cryo-TEM imaging where a similar coating was observed.

EDX analysis was carried out in order to identify the elements present in the observed coating. The EDX spectrum contained the expected Ba and Ti peaks present due to the composition of the nanoparticles themselves and a Cu signal in accordance with the use of a Cu TEM grid. In addition there were clear Ca, P, Na, Cl and Mg peaks (Figure 2). All these elements can be found within the cell culture media used which contains, inorganic salts, amino acids and vitamins, in addition to proteins from the FBS supplement. From the EDX maps, these elements appeared to be accumulating around

IOP Conf. Series: Journal of Physics: Conf. Series **902** (2017) 012006 doi:10.1088/1742-6596/902/1/012006

the BT nanoparticles forming the coating seen in the TEM images. The peaks for Ca and P were the most prominent which suggested that the coating was dominated by these two elements.

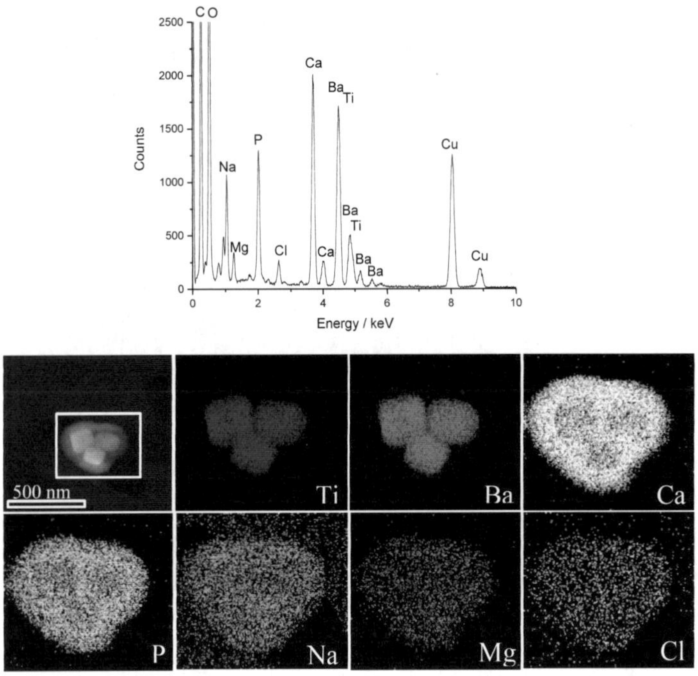

Figure 2: STEM-EDX data of DC BT in CCCM. In the EDX spectrum Ca and P are the most prominent peaks but also present were Cl, Mg and Na, all of which spatially match to the coating imaged around the BT nanoparticles.

Cryo-TEM of the suspension permits 'native' state imaging and analysis and confirmed the presence of the coating observed in TEM of the DC sample (Figure 1(b)). This suggested that it was not a result of any drying artefacts common to the DC technique but was present in suspension. Cryo-STEM-EDX analysis, carried out whilst maintaining the vitreous ice surrounding the specimen, showed a significantly different composition to the coating compared to that obtained from TEM of DC BT in CCCM (Figure 2). As expected O, Ba, Ti and Cu peaks were again present. So too were Ca and P peaks, however, the other elements observed in TEM of the DC sample (Na, Cl and Mg) were far less prominent; only a Cl peak could be confidently distinguished from the spectral background (Figure 3).

We suggest that this compositional difference in the coating can be explained by considering that during the process of plunge freezing, the nanoparticles were maintained in suspension whereas the DC sample preparation caused additional components of the CCCM to accumulate around the BT nanoparticles during drying. Thus, the cryo-STEM-EDX data suggested a coating did indeed form around the BT nanoparticles in suspension and that this was entirely Ca and P rich. This confirmed a significant interaction occurred between the BT nanoparticles and CCCM which is presumably driven by the pH of the CCCM which influences the surface charge of the nanoparticles and in turn causes interactions with certain ions in the media.

The smaller signals of Ca and P relative to Ba and Ti seen in cryo-STEM-EDX could be due to the thinner coating that was observed in cryo-TEM imaging compared to DC TEM imaging (Figure 1). However, considering the lack of any Mg and Na signals this confirmed that these elements did not adsorb to the nanoparticle surface. Although there was still a signal in the cryo-EDX spectrum for Cl, when looking at the EDX map, a far more dispersed distribution of Cl across the analysed area was seen compared to the EDX Cl map from DC TEM (Figure 2). This was indicative of the element simply being present in suspension and not as part of the observed coating.

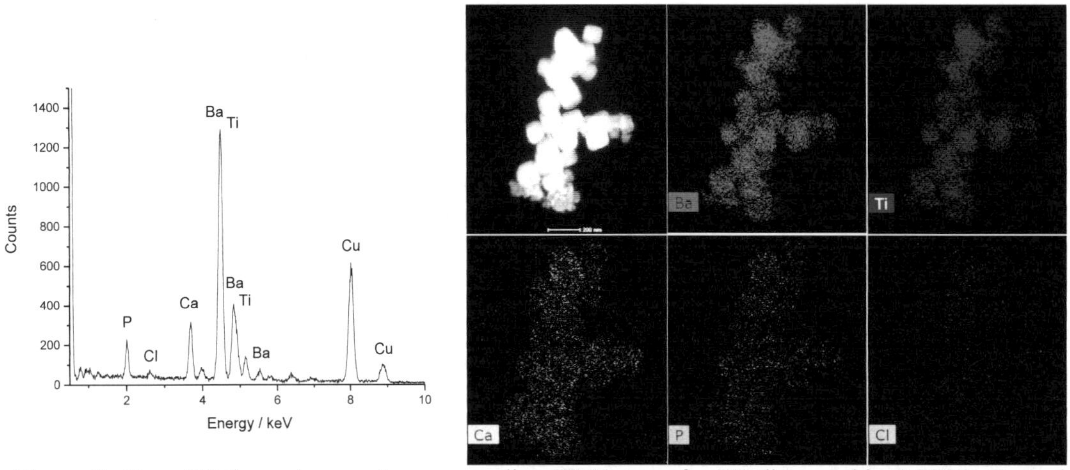

Figure 3: Cryo-EDX spectrum and corresponding EDX maps for BT dispersed in CCCM. Ca and P are present in the spectrum but through cryo analysis no visible Na and Mg peaks were present and the Cl peak was only just distinguishable. Analysis was carried out using a probe current of 40 pA.

Considering the importance of Ca and P in cell function it is likely that the accumulation of these elements around BT nanoparticles in CCCM will significantly influence the uptake and impact of the nanoparticles during *in vitro* cellular exposure carried out using DMEM. In view of this, further concerns arise due to the variable composition of commercial cell culture media; for example Roswell Park Memorial Institute, another common culture, contains roughly six times as much phosphate in comparison to DMEM [5]. Thus it would be expected that the behaviour of a nanoparticle system within one cell culture media may not be the same as within another.

4. Conclusion

A significant variation between STEM-EDX spectroscopy carried out on DC and plunge frozen samples of BT nanoparticles in CCCM established that drying artefacts in DC TEM can dramatically alter the perceived nano-bio interactions. Cryo-TEM analysis was successfully carried out to enable identification and elimination of detrimental drying artefacts to reveal a significant Ca and P rich coating formed around BT nanoparticles when dispersed in CCCM.

The study has identified the pressing need to continue the development of analytical cryo-TEM in order to more accurately characterise nanoparticles within complex media and confirmed cryo-STEM-EDX as one way to do this. Future work will look to monitor the nano-bio interactions of other nanoparticles systems within CCCM and continue to develop the possibilities within cryo-analytical TEM whereby cryo-EELS and low dose imaging can be performed.

References

[1] Shi J, Kantoff P W, Wooster R and Farokhzad O C, *Nat. Rev. Cancer.* 2017 **17** 20-37.
[2] Genchi G G, Marino A, Rocca A, Mattoli V and Ciofani G, *Nanotechnology* 2016, **27**.
[3] Eckert J O, Hung-Houston C C, Gersten B L, Lencka M M and Riman R E, *J. Am. Cer. Soc.* 1996 **79** 2929-39.
[4] Matar O, Posada O M, Hondow N S, Wälti C, Saunders M, Murray C A, Brydson R M D, Milne S J and Brown A P, *J. Phys. Conf. Ser.* 2015 **644** 012037.
[5] Moore T L, Rodriguez-Lorenzo L, Hirsch V, Balog S, Urban D, Jud C, Rothen-Rutishauser B, Lattuada M and Petri-Fink A, *Chem. Soc. Rev.* 2015 **44** 6287-6305.

IOP Conf. Series: Journal of Physics: Conf. Series **902** (2017) 012007 doi:10.1088/1742-6596/902/1/012007

Kinetically controlled fabrication of gold nanorods and investigation of their thermal stability via in-situ TEM heating

N. Chankhunthod[1], Z. Aslam[1], K. Critchley[2], S. D. Evans[2], and R. Brydson[1]
[1] School of Chemical & Process Engineering, University of Leeds, Leeds, LS29JT, UK
[2] School of Physics & Astronomy, University of Leeds, Leeds, LS2 9JT, UK

E-mail: ml13n2c@leeds.ac.uk

Abstract. Size controlled CTAB-capped AuNRs with various aspect ratios (ARs) ranging from 1.63±0.13 to 4.12±0.25 were synthesized following a modified seed-mediated method. Their thermal stability was examined by in-situ TEM heating. The results revealed a structural change from rods to spheres with increasing temperature. At lower temperatures < 300°C, the AR decreased rapidly and, for temperatures >600°C, particles became increasingly spherical. This behaviour occurred at temperatures lower than the melting point of bulk gold supporting a surface diffusion mechanism with material diffusing from the tips and redepositing at the middle of the rods. The rate of change in AR appeared to increase for thinner AuNRs.

1. Introduction

Gold nanorods (AuNRs) play an important role in various applications due to their shape-dependent optical properties which can be tuned into the near infrared (NIR) region for the purposes of in-vivo photothermal therapy and ultrasound imaging which take advantage of the NIR window in biological tissue [1]. AuNRs exhibit two surface plasmon resonances: transverse and longitudinal [1, 2]. The longitudinal plasmon resonance has been found to be very sensitive to the aspect ratio (AR) of the rods, thus controlling the AR via a facile synthesis procedure is of importance. However, there are reports of a decrease in the AR of AuNRs during exposure to ultrafast pulsed laser radiation and thermal heating [3]. Hence the thermal stability study of AuNRs is of interest both fundamentally and for their use in specific medical applications.

2. Experimental Procedure

2.1 Synthesis of CTAB-capped AuNRs

In this work, a seed mediated method following Nikoobakht and El-Sayed [4] was used to obtain homogeneous AuNRs with various ARs; kinetic control via pH adjustment was introduced in order to slow down the growth rate and achieve a homogenous distribution of AuNRs. The synthesis involved 2 main steps: seed growth and growth in solution described below.

1.) *Seed solution,* 5 mL of 0.2 M CTAB was gently mixed with 5 mL of 0.5 mM of $HAuCl_4$. Then, 0.6 mL of 2 hr aged ice-cold 0.01 M $NaBH_4$ was rapidly injected and mixed by shaking. The solution was left for 2 hr.
2.) *Growth solution,* 10 mL of growth solution containing 5 mL of 0.2 M CTAB and 5 mL of 1 mM $HAuCl_4$ was first prepared in 5 different vessels. Then, 4 mM of $AgNO_3$ was added in differing amounts (150, 200, 250, 300, 350 μL) in each vessel and gently mixed in order to obtain different AR rods (denoted samples 1-5 respectively). Kinetic control was achieved via addition of 12 μL of 12 M HCl and gentle mixing. Next, 70 μL of 0.0788 M of ascorbic acid was mixed into the growth solution resulting in a colourless appearance. The final step was to

rapidly inject 12 μL of seed solution into the growth solution without any mixing. This was incubated at 30°C for 5 hr.

2.2 Materials Characterization and TEM heating experiment

AuNR samples were initially centrifuged twice and redispersed with the same amount of Milli-Q water to remove the excess CTAB. They were then characterized by UV-vis spectroscopy, and TEM. 1 mL of AuNRs was put in a UV-vis cuvette and examined in a Perkin-Elmer lambda35 model spectrophotometer to obtain the optical absorption spectrum from 400 to 900 nm. All TEM images were obtained in bright field mode from a FEI Tecnai G2 TEM/STEM operating at 200 kV. TEM samples were prepared by dropping 2 μL of AuNRs suspension onto holey carbon film 400 mesh copper TEM grids and naturally dried at room temperature. The thermal stability of AuNRs was studied via in-situ TEM. The samples were prepared by dropping 2 μl of centrifuged AuNRs onto a Si_3N_4 heating chip (DENS solutions) which was then allowed to dry. The chip was loaded into the in-situ DENS heating holder and loaded into the TEM. The chip was heated from room temperature up to 1000ºC, applying a 50ºC increase every 5 minutes. The morphology of the rods in the centre of the heating chip were monitored.

3. Results and discussion

3.1 Synthesised CTAB-capped AuNRs

The kinetically controlled modified synthesis method of CTAB-capped AuNRs produced well-controlled and reproducible AuNRs with various ARs. The colour of the dispersion gradually changed over 20-30 min depending on the aspect ratio of the rods. Figure 1(a) illustrates the colour differences of CTAB-capped AuNRs and their corresponding UV-vis spectra. The samples exhibited a red-shift of the absorption spectra as a function of increasing AR. The longitudinal plasmon resonances of the synthesized rods ranged from the visible region (575 nm) to the NIR (800 nm) according to their AR. Statistical measurements of particle size (200 particles per sample) from TEM images were undertaken using ImageJ software. The results showed that the AR of AuNRs increased as the amount of $AgNO_3$ employed in the synthesis increased. The ARs of AuNRs were 1.63±0.13, 2.14±0.22, 2.92±0.25, 3.43±0.32, 4.12±0.25 for samples 1-5 respectively, as seen in figure 1(b). This shows the longitudinal plasmon mode is very sensitive to the AR, with only a small change in resulting in a significant shift of the absorption spectra.

The absorption spectra also exhibited a transverse plasmon resonance band located at around 520 nm, which did not vary significantly between different samples, in agreement with literature. This transverse absorption mode represents the oscillation of electrons over the short axis of the rods and the absorption maximum was similar to that of spherical particles which had a peak resonance at around 520 nm. However, in sample 5, the transverse peak (orange colour) was broad indicating that various sizes of spherical particles were produced. The presence of AuNRs were confirmed by HRTEM and selected area electron diffraction (SAED). An SAED pattern of a group of AuNRs (inset 2(a)) showed fcc Au lattice reflections of (111), (200), (220), and (311). In figure 2(b), HRTEM showed a lattice spacing of 0.148 nm along the rod axis which can be assigned to the (220) spacing of fcc Au.

3.2 TEM heating experiment

AuNRs with different ARs were investigated by in-situ TEM heating. The results illustrated a structural change of AuNRs with increasing temperature, as shown in figure 3. The aspect ratio clearly started changing, albeit with only a small decrease, at relatively low temperatures ca. 100ºC. More significant changes were observed at higher temperatures, with the rod length decreasing and the rod width increasing. The rate of change of AR, the slope of the graph in figure 3(b), was large for temperatures up to 300ºC and then reduced and became relatively constant at higher temperatures. For temperatures >600ºC, particles became increasingly spherical for all rods monitored. There were some variation in results with some rods exhibiting a significant change in AR at high temperatures, e.g. < 700ºC for rods 1, 2 and 3.

Longer rods were found to be more unstable and exhibited a higher rate of change of AR with temperature than shorter rods. The change in AR depended on the surface area to volume ratio [5]; thinner rods (rods 1-4) exhibited a more rapid decrease in AR as compared to thicker rods (rods 5-7). For rods with different lengths but the same width, the longer rods showed more a significant change with temperature than the shorter rods. Interestingly for rod 4, which is thinnest rod with an AR~3.82(L/W;28.34/7.41nm), the surface of AuNR started melting and diffused from the tip to the middle of the rod at ca. 150°C. At 250°C, a significant decrease in length was observed. At higher temperatures the rate of change increased and the rod transformed into a sphere at about 800°C. Initial room temperature TEM images were compared with those at the end of heating profile to calculate the particle volume change. Assuming a cylindrical rod with two hemi-spherical endcaps, the estimated initial volume was 1180.29 ± 0.09 nm^3 whilst the spherical particle volume after heating was 1134.98 ± 0.05 nm^3. This perhaps indicates a slight change in volume which may occur during to surface melting via evaporation or diffusion onto the support film. The thermal transformation mechanism of rods was revealed by heating at a lower temperature for a longer period of time (figure 4). This revealed evidence for surface melting and surface diffusion at temperatures lower than the melting point of bulk gold arising from the minimization of surface energy [6]. The tips of rods, with a large radius of curvature and hence higher surface energy, were most reactive with material diffusing and redepositing towards the middle of the rods, resulting in a decrease in aspect ratio.

Overall the studies revealed that rods with aspect ratios of about 4.0 - 4.5, with a surface plasmon resonance lying at about 800 nm, would be a good candidate for nanomedical applications, both in terms of its optical properties and thermal stability; the useful ranges of temperatures, whereby heating did not significantly change optical properties, were up to 100 °C.

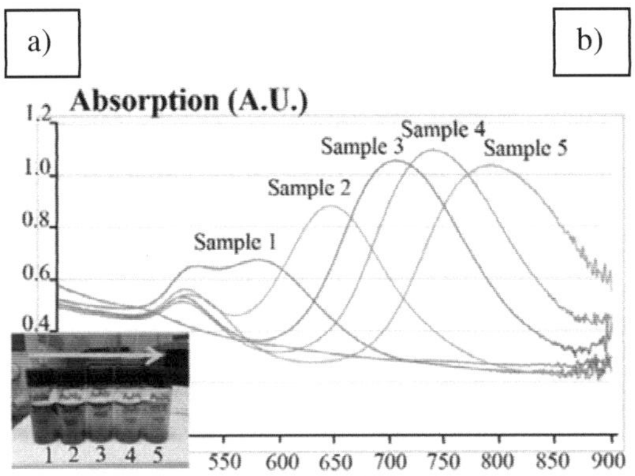

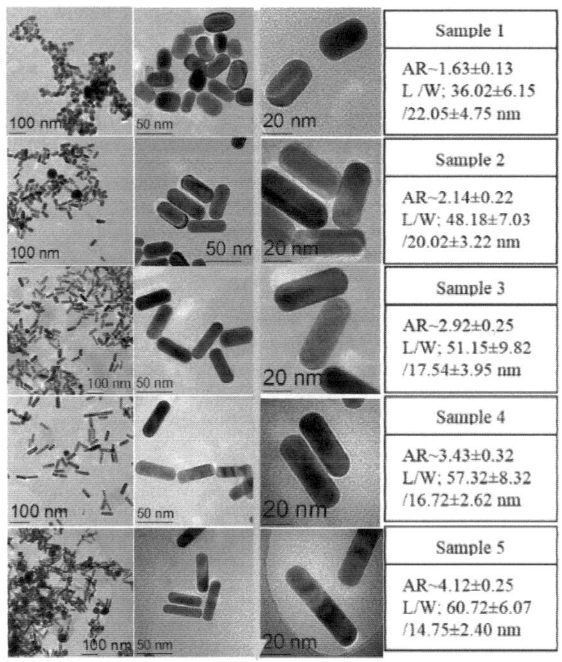

Figure 1. a) Absorption spectra of CTAB-capped AuNRs with various ARs. Inset shows the change in colour of as-synthesized AuNRs from purple to pink indicating the increase in rod AR; b) TEM images of CTAB-capped AuNRs with various ARs from 1.63±0.13 to 4.12±0.25

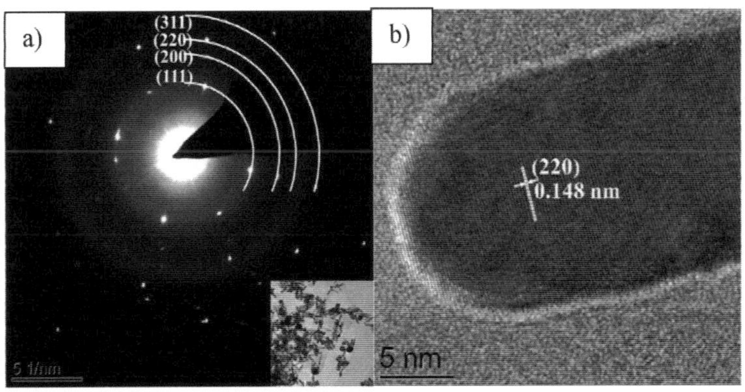

Figure 2. a) Electron diffraction pattern of a group of AuNRs shown in the image inset; b) HRTEM image of AuNRs, showing [220] direction along the rod axis

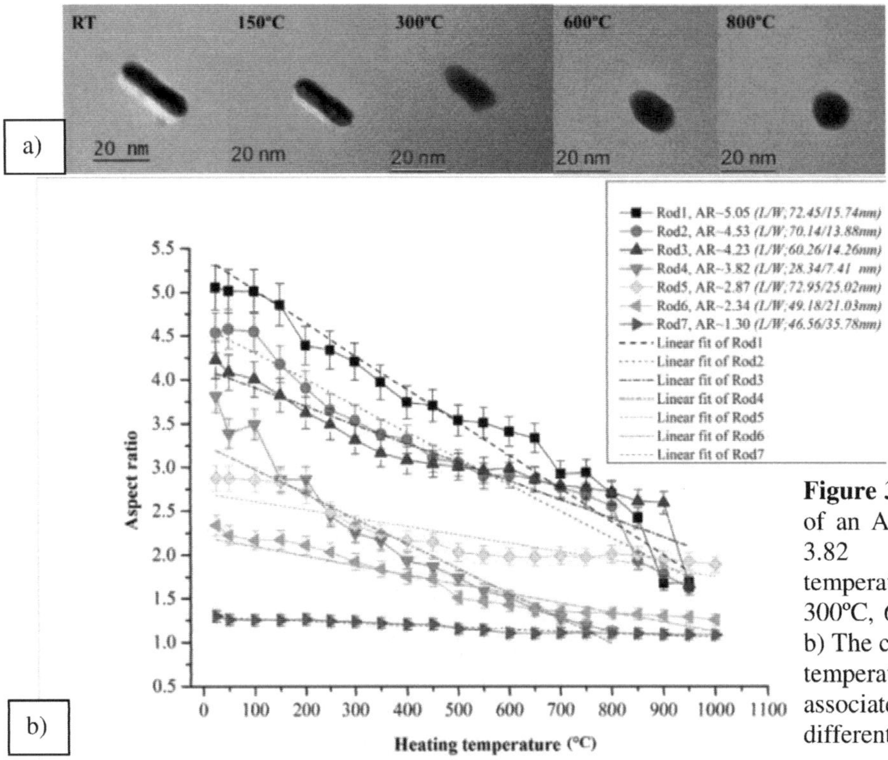

Figure 3. a) TEM images of an AuNR with AR of 3.82 at different temperatures: RT, 150°C, 300°C, 600°C and 800°C. b) The change in AR with temperature and associated linear fits for different AR rods

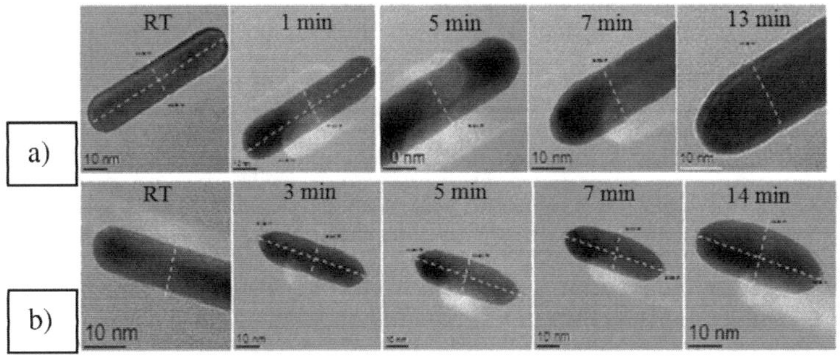

Figure 4. Low temperature heating experiment: TEM images of an AuNR at a) 100°C and b) 150°C for 15 mins, illustrating the transformation mechanism arising from surface diffusion from tip and redeposition at the middle of the rods

4. Conclusion

Size controlled AuNRs with various ARs were reproducibly synthesised following a modified seed-mediated method together with pH adjustment. After heat treatment via in-situ TEM heating, there was a structural change from rods to spherical particles which depended on the exact rod morphology. These findings are important for the application of AuNRs in photothermal therapy and ultrasound imaging, which require no change in optical properties during heating. This study attempts to define a usable temperature range for AuNRs in such applications which could be then simulated by modelling for a given laser power.

References

[1] Chen, H., et al., Chemical Society Reviews, 2013. **42**(7): p. 2679-2724.
[2] Cao, J., T. Sun, and K.T.V. Grattan, Sensors and Actuators B: Chemical, 2014. **195**: p. 332-351.
[3] Petrova, H., et al., Physical Chemistry Chemical Physics, 2006. **8**(7): p. 814-821
[4] Nikoobakht, B. and M.A. El-Sayed, Chemistry of Materials, 2003. **15**(10): p. 1957-1962.
[5] Jana, N.R., et al., Langmuir, 2002. **18**(3): p. 922-927
[6] Zou, R., et al., Colloids and Surfaces A: Physicochemical and Engineering Aspects, 2010. **372**(1–3): p. 177-181

Coalescence dynamics of size-selected gold clusters studied by time-resolved transmission electron microscopy

J. Liu [1], D. Foster [1], Z.Y. Li* [1], N. Wilkinson [2] and J. Yuan [3]

[1] School of Physics and Astronomy, University of Birmingham, Birmingham B15 2TT, U.K.
[2] Gatan UK, 25 Nuffield Way, Abingdon, Oxon. OX14 1RL, U.K.
[3] Department of Physics, University of York, York YO10 5DD, U.K.

z.li@bham.ac.uk

Abstract. Coalescence dynamics of size-selected gold (Au) clusters (each with nominal 923 atoms), on amorphous Si_3N_4 substrate at room temperature, has been studied via time-resolved transmission electron microscopy (TEM). We found that the clusters approached each other in two stages. In the first stage, the drift velocity was independent of the particle separation and could be attributed to beam-induced random motion. In the second stage, the clusters were found to jump into contact with a much higher final averaged speed. This is independent of beam dose rates and is attributed to the van der Waal attraction.

1. Introduction

Coalescence of supported metal nanoparticles/clusters is a common phenomenon in the field of catalysis, whose process is known to relate to many factors, such as size, structure, orientation, temperature, support and environment [1, 2]. It is critical to quantify these factors for understanding the physical mechanism of the coalescence if one wants to rationally design a catalyst system. Previous experimental studies of nanoparticle coalescence are mainly on particles of different sizes [1, 3], thus it is difficult to differentiate the effect of size as to other factors such as temperature, structure, etc. Here we utilised the unique capability of an atomically size-controlled cluster source in our group, together with a fast acquisition feature of a microscope camera, to study the coalescence phenomena between gold (Au) clusters. Specifically, we reported the effects of the cluster size and electron dose on the coalescence of Au clusters.

2. Experimental details

The size-selected Au clusters were produced in a magnetron-sputtering gas condensation cluster source equipped with a lateral time-of-flight mass filter [4,5]. In this study, the clusters with atom number of 923±23 (the resolution of the mass filter was ±2.5%) were deposited at 1 eV/atom onto an amorphous Si_3N_4 substrate at room temperature. The dynamics of cluster coalescence at room temperature was captured using a Gatan OneView camera in a 200 kV JEOL 2100F transmission electron microscope (TEM) and at an imaging rate of 302 frame/sec. The videos were recorded under electron dose rate of 2.9×10^5, 5.1×10^5 and 6.5×10^5 $e^-/nm^2/s$, respectively, while keeping the other imaging conditions, such as magnification, frame rate, etc. the same.

3. Results and discussions

Figure 1(a) shows a TEM image example of the deposited Au clusters on an amorphous Si_3N_4 substrate. The diameter distribution histogram, Figure 1(b), shows three discrete peaks at 2.73 ± 0.13 nm, 3.66 ± 0.12 nm and 4.26 ± 0.41 nm, respectively. This size distribution is different from evaporated Au nanoparticles, which had a single broad distribution peak [6]. The Au-Au nearest neighbour distance is 0.288 nm [7] and the theoretical diameter of a spherical Au_{923} cluster (full-shell geometry) is 3.45 nm, which is close to the value of 3.36 ± 0.16 nm measured by an aberration-corrected scanning transmission electron microscope (AC-STEM) [8]. These values are over 0.6 nm bigger than the value measured here by TEM. This is most likely due to the difficulty of visualisation of the outermost layer of the Au cluster in the aberration un-corrected TEM image because of the Fresnel effect. Therefore, the clusters with an average diameter of 2.73 ± 0.13 nm are assigned as single Au_{923} clusters. The diameter (D in nm) for a 2× or 3× Au_{923} cluster (assuming a spherical shape) is calculated with the following formula,

$$D = 3.36 * \left(V/V_{Au_{923}}\right)^{1/3} - 0.576$$

where 3.36 nm is the average diameter of Au_{923} clusters measured by an AC-STEM, V is the volume of a cluster, $V_{Au_{923}}$ is the volume of a Au_{923} cluster, and 0.576 nm is twice of the Au-Au nearest neighbour distance. Here, the volume ratio is either 2 or 3, then the calculated diameter of a 2× and 3× Au_{923} cluster is 3.66 nm and 4.27 nm, respectively. Thus, we attribute the next two peaks in Figure 1(b) to the clusters merged with two and three Au_{923} clusters, respectively, with the merged clusters are approximated spherical as well. We notice that a slightly larger proportion of clusters has size bigger than Au_{923} on the current amorphous Si_3N_4 than that found on amorphous carbon, suggesting a weaker interfacial adhesion of Au_{923} to the current Si_3N_4 substrate.

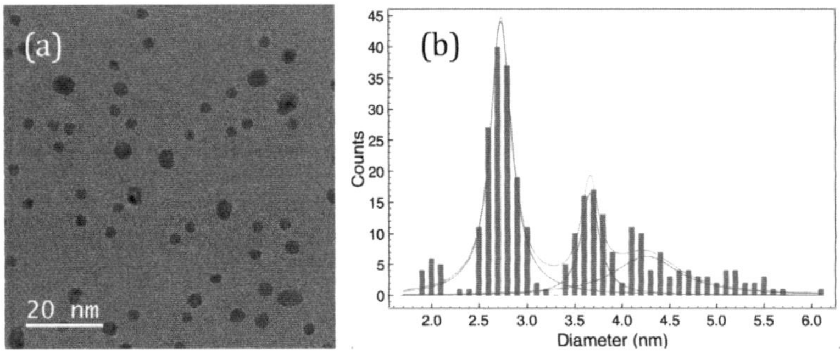

Figure 1 (a) A typical TEM image of the size-selected Au_{923} clusters deposited onto an amorphous Si_3N_4 substrate. (b) Diameter histogram of the Au clusters. The solid line curves are the Gaussian fitting.

Under the electron beam exposure, the clusters initially stayed still at room temperature. As the beam dose rate increased to the order of 10^5 $e^-/nm^2/s$, they started to move in random directions on the surface, indicating that this Brownian-like motion was largely electron beam driven. The e-beam driven phenomena are commonly observed in electron microscopy studies, particularly for nanomaterials, for example, ref. [9], Despite that, a recent work [10] suggests that an electron beam driven stochastic motion can be very useful to study the mechanism of thermal driven stochastic motion underpinning coalescence in catalytic particles. Here, we focus on those clusters that got closer and coalesced. As demonstrated in Figure 2(a) (bottom left), the two Au_{923} clusters approached towards each other when imaged at an electron dose rate of 6.5×10^5 $e^-/nm^2/s$ (the first two images). When they contacted, a "neck" was seen to form and widen at the contacting planes (the third and fourth images), and finally the two clusters merged into one as shown in the last image.

The distance between the geometric centers of the two clusters was measured and the relative speed was calculated from the distance changing between frames, as shown in Figure 2 (b) and (c). It was noticed that there was a steady fluctuation in the centre-to-centre distance followed by a big change of about 1 nm (indicated by an arrow) at about 0.3 s, just before the two clusters merged. The average approaching speed, before the clusters coalesce, is 48 ± 36 nm/s. This speed is relatively slow compared to the contacting speed in the final approach at about 320 nm/s. It suggests that the approaching process consists a long-range drift superimposed on the stochastic motion of the clusters. It also shows that a fast camera is essential to resolve the details of this final contacting process, since only one or two frames were captured even at the frame rate of 302 frames/s.

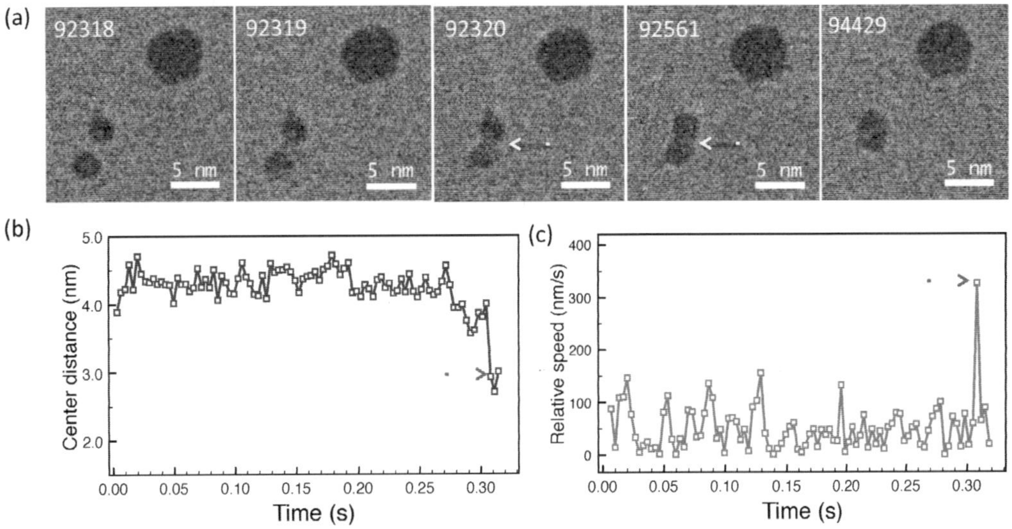

Figure 2 (a) Coalescence between two Au_{923} clusters (bottom left). The frame number is labelled on the top left corner of each image and the time interval between each frame is 3.31 ms. The yellow arrows indicate the "neck" formation. (b) Center-to-center distance versus time, and (c) relative speed changing of the two clusters versus time. The distance and relative speed were measured from frame 92220 to 92320. The contacting points are indicated by the green arrows.

It is interesting to compare our experimental result with available simulations. Wang *et al.* [11] studied the coalescence process of two 8 nm diameter gold nanoparticles on a solid silicon surface by conducting Molecule Dynamic (MD) simulations. They defined the approach as the stage before two particles were in contact and the coalescence as the structure relaxation after the contact. The interactions of these two processes were simulated using Lennard-Jones potential and Embedded Atom Model (EAM), respectively. They found that the coalescence was not obvious when the temperature was lower than 800 K. This is consistent with our observation at room temperature (~300 K) that the clusters were largely immobile at low beam intensity. Focusing on the approaching process, the simulation (which was conducted over 0.2 ns time window) showed that the average speed was 0.7 nm/ns at 300 K. However even within this simulation window, the actual speed increased rapidly as the particle separation decreased because of the van der Waal attraction. Thus, one may be sure that the final jump distance (< 0.14 nm) would be completed within the sub-nanosecond time scale. If the same jumping event was recorded with a 3.31 ms frame interval as in the present experimental study, the corresponding 'frame-average' approach speed would be about 40 nm/s. This is about the same order of magnitude as our experimental results, Figure 3(c). However, we also notice that the simulation was for 8 nm Au nanoparticles on crystalline Si substrate and our present experiments were conducted on 3 nm Au clusters on an amorphous Si_3N_4 substrate. Therefore, more realistic simulations would be useful in order to properly interpret the experimental results. Nevertheless, the order-of-magnitude agreement may not be surprising as gold atoms interact weakly with both substrates and as the gold particles retained their spherical shapes in both simulation and experiments.

To study the effect of the electron beam irradiation, we analysed the experimental 'frame-averaged' contacting speed under different electron dose rates (Figure 3). We found that the average contacting speed varied from about 100 to 400 nm/s and was almost independent of the electron dose rate employed. This suggests again that the rate of the final approach is not due to the beam-induced effect. In addition to the van der Waal attraction, other factors, such as interaction with the substrate and crystalline structure of nanoclusters [3], may also contribute to the average contacting speed.

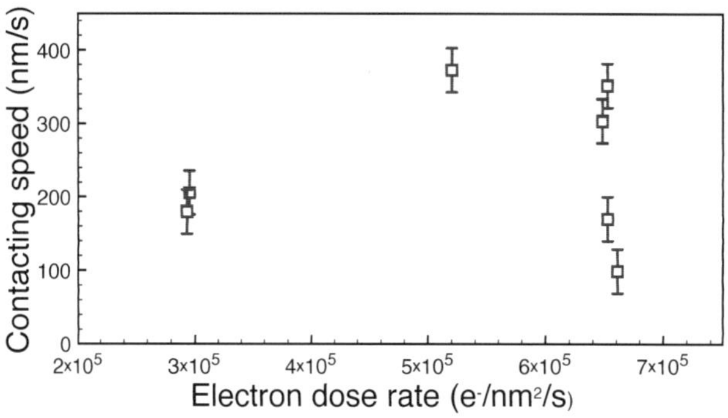

Figure 3 Contacting speed against electron dose rate. The error bar is calculated from the distance measurement error of 0.1 nm.

4. Conclusions

In summary, we have studied the coalescence dynamics of size-selected Au_{923} clusters by using a time-resolved TEM. The ultra fast camera allows us to estimate the timescale for the contact process to be millisecond or less. The final contacting speed of two clusters ranges from about 100 to 400 nm/s and is weakly dependent of the electron doses employed. The variation of contacting speed under the same irradiation condition indicates that other factors, such as orientation and substrate, may also play a role. The work illustrates the usefulness of time-resolved high-resolution TEM in combination of size-selected atomic clusters, in studying the atomistic process underlying the fundamental process of cluster coalescence with implications for future rational design of catalytic nanoparticles. It also highlights the need for joint theoretical and experimental studies over the same system.

Acknowledgement

We thank Dr. Yu-Lung Chiu for his assistance in using the microscope facilities at Centre of Electron Microscope for conducting this work and Gatan for the loan of an one-view camera.

References

[1] Jose-Yacaman M, Gutierrez-Wing C, Miki M, Yang D Q, et al. 2005 *J Phys Chem B* **109** 9703
[2] Barnard A S 2010 *Rep Prog Phys* **73** 086502
[3] Grammatikopoulos P, Cassidy C, Singh V, Sowwan M 2014 *Sci Rep* **4** 5779
[4] Li Z Y, Young N P, Di Vece M, Palomba S, et al. 2008 *Nature* **451** 46
[5] Von Issendorff B and Palmer R E 1999 *Rev Sci Instrum* **70** 4497.
[6] Young NP, Li Z Y, Chen Y, Palomba S, et al. 2008 *Phys Rev Lett* **101** 246103
[7] Kittel C 1986 John Wiley & Sons, Inc.: New York
[8] Hu K J, Plant S R, Ellis P R, Brown C M, et al. 2015 *J Am Chem Soc* **137** 15161
[9] Zhang D, Jin C, Tian H, Xiong Y, et al, 2017 *Nanoscale* **9** 6327
[10] Hong J, Pan H, Hu Z, Lv D, et al. 2017 *Nano Lett* **17** 3383
[11] Wang J, Chen S, Cui K, Li D and Chen D 2016 *ACS Nano* **10** 289

Electron Beam Transformation of Glass Nanoparticles

M. M. Sabri[1,2], G. Möbus[1]

[1]Dept of Materials Science and Engineering, University of Sheffield, Sheffield, S1 3JD, UK
[2]Present Address: Department of Physics, Faculty of Science and Health, Koya University, University Park, Kurdistan Region, Iraq
g.moebus@sheffield.ac.uk

Abstract. The phenomenon of electron irradiation induced quasi-fluid flow in solid amorphous silicates is examined. High-intensity electron beam irradiation of glass nanoparticles in TEM leads to a variety of transformations including rounding, bead formation, and phase separation. Evidence for possible rise in temperature, charging fields, and enhanced diffusion is collected and compared, while the latter effect (loosening of glass network) is shown to dominate.

1. Introduction

The deliberate or accidental irradiation of amorphous solid oxide materials has attracted major interest in a variety of research fields, including glass patterning for e.g. wave guide applications, modulated index or photochromic glass fabrication, nanoplasmonic glass/metal nanoparticle composites, and for radioactive material storage and disposal by vitrification [1]. While laser, ion beam, and radioactive radiation sources have been widely used in the past to study glass-irradiation behaviour, we concentrate here on electron beam irradiation. Expected effects of the irradiation of an initially homogeneous glass would include: density changes (either up or down), diffusivity changes, phase separation, and precipitation of crystalline particles.

Relevant recent research findings include the *in-situ* TEM nanomechanical testing of irradiated glass with ultra-soft plastic flow [2], point defects and high diffusion rates [3], while our preliminary work has found evidence for surface shape changing and pseudo-ductile behaviour of glasses in TEM [4]. The finding that "solid", normally brittle, material appears mechanically soft when objects are of nanoscale dimensions and irradiated at high enough electron intensity, is often referred to as "Radiation Induced Fluidity" (RIF) or quasi-melting.

Here we aim at collecting evidence which could help decide between three basic mechanisms possibly underlying the RIF-observations of fine glass particles in TEM:

M1: The first possible option for mechanisms is the thermal "melting"-like transformation. Electron-beam induced sample heating could lift the particle to either above glass-transition, or further up to near melting temperature. The result would be some liquid-like glass behaviour with rough surfaces turning smooth and approaching overall spherical or at least round shapes by surface tension.

M2: The second option is charging and related ionic migration and atomic ablation due to local electric fields of varying strength [5, 6]. Diffusion and Ablation rates would follow local field strength and contribute to enhanced flow and material removal (the latter preferentially at sharp corners with peaks in field strength)

M3: Radiation-induced fluidity (RIF) in the narrower meaning [2,3] involves neither change in temperature nor de-homogenisation by ion migration, but merely relies on high point defect generation rates triggering high diffusion rates, and allowing the glass particles to reconfigure according to surface tension "as if they were liquid-like", with a massively softened glass network.

In continuation of [4, 7-9], we present examples supporting or marginalising each of the mechanisms M1-M3, illustrated in Figure 1:
(a) centripetal surface tension with sphere formation under volume conservation,
(b) electric field enhancement at corners and ablation leading to rounding under volume loss;
(c) surface diffusion leading to rounding with overall shape conservation.

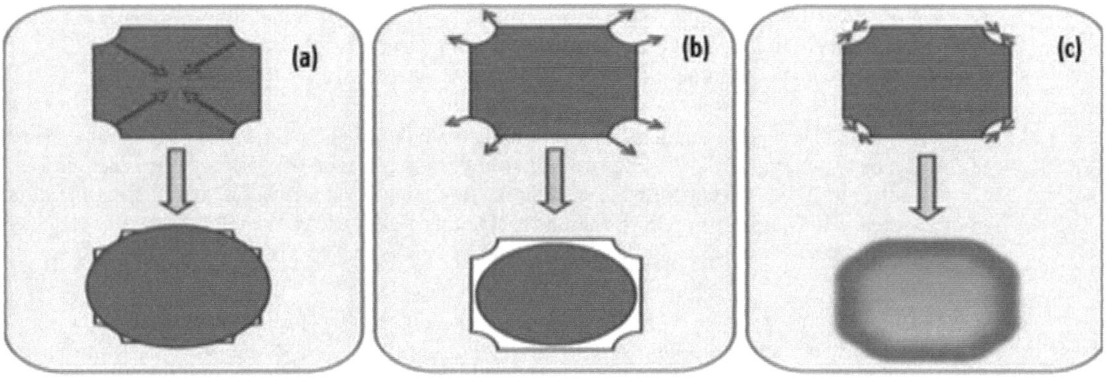

Figure 1. Schematic of 3 possible glass particle transformation mechanisms by electron irradiation.

2. Materials and Methods

Alkali-borosilicate glasses have been synthesised with varying boron content and with a variety of extra cations loaded, as part of earlier precipitation studies. Melting at 1300-1400° C followed by quenching in water generates a glass-frit which is ground into small micron or nano-sized fragments by pestle-and-mortar and suspended onto Cu-grid/C-film TEM specimens. Microscopy is performed at 200 or 300 kV (JEM 2010F or JEM 3010) with intensities varied via condenser focusing, normally at largest aperture. Occasionally very high intensities are achieved by removing any condenser aperture.

Glasses used include in particular:

Table: Chemical compositions of glass (mol.%).

	Compositions (mol.%)								
Glass	B_2O_3	SiO_2	Na_2O	Li_2O	CeO_2	Cr_2O_3	ZrO_2	CuO	ZnO
NBS	20	63.4	16.6	0	0	0	0	0	0
NLBS-CCZ	25.7	51.4	8.6	4.3	4	2	4	0	0
Cu-NBS	15	50	15	0	0	0	0	20	0
Zn-BS	20	20	0	0	0	0	0	0	60

3. Results

3.1. Slow particle rounding

In Figure 2 (a, b) an original, randomly shaped, glass particle with sharp corners is irradiated and its transformation to round shape followed. This slow change is well controllable, and while the length of the anisotropic particle is clearly shrinking, the short aspect is growing. The latter points to volume conservation to a good degree, and proves materials flow. The alternative rounding mechanism by preferential ablation only can be excluded here, as it would result in a smaller particle the footprint of which fits completely into the original shape. No signs of glass de-homogenisation (phase separation or precipitation) are seen, all evidence supports mechanism M3.

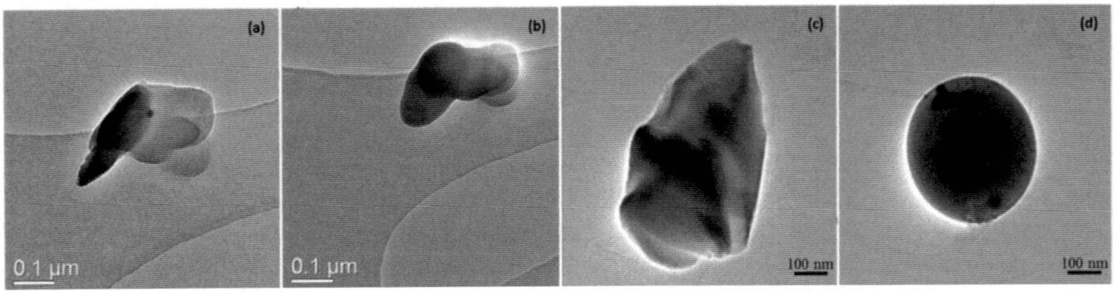

Figure 2. Example of slowly transformed (a, b) glass fragment (NBS-glass; JEM 2010F at 200 kV), and comparison to (c, d) rapidly transformed fragment (Cu-NBS glass [8]; JEM 3010 at 300 kV).

3.2. Fast bead formation.
The 2nd particle (Figure 2 (c, d)) shows distinctively different behaviour, as here a rapid not well controlled one-step transformation from the original rough and non-symmetric shape into a sphere is observed. While the compositions of the two glasses are not different enough to explain the behaviour, two details matter: The bead-formation is only observed for particles a few 100-nm large, and there is also a particularly sparse carbon-lace support visible at the particle location. Both details would trigger a much larger jump in temperature compared to Figure 2 (a, b), and might support a combination of mechanisms M1 and M3.

3.3. De-homogenisation.
Phase separation, particle precipitation and shape-rounding can all coincide, see Figure 3 (a, b). While only observed occasionally, the effect supports increased diffusion and assures glass homogeneity is preserved on the >10 nm scale (the composition near the corners is identical to the fragment centre), again supporting mechanism M3. Yet, electric field induced ion migration and ablation (M2) play a complementary role as evidenced in Figure 3 (c, d). Here, unlike in any other figure, the beam diameter used for irradiation was smaller than the glass fragment. Therefore, local charge-patterns and radial field distributions are generated. The response by the glass showing enhanced ablation (volume loss) and formation of two cation-enriched rings indicates mechanism M2 is dominant.

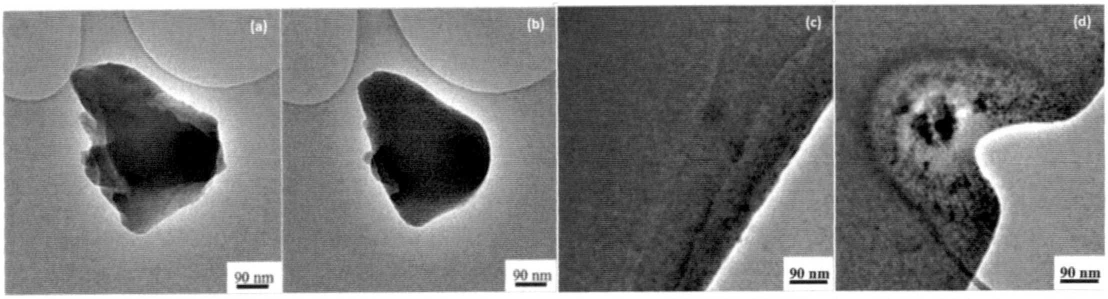

Figure 3. (a, b) Particle rounding with phase separation (NLBS-CCZ). (c, d) Field-induced ion migration alongside materials ablation and precipitation. Cu-NBS glass [8], JEM 3010; 300 kV.

3.4. Heating holder TEM.
Glass irradiation has been performed on a Zn-borosilicate glass [9], initially generating many Zn-nanoparticles (Figure 4 (a, b)) and some volume shrinkage. Subsequent heating up to 300° C in situ triggered changes in microstructure such as precipitation growth and particle-coagulation, while

intensities were kept low for imaging. However, no rounding or shape change due to heating is observed, and nearby un-irradiated particles did show neither rounding nor precipitation.

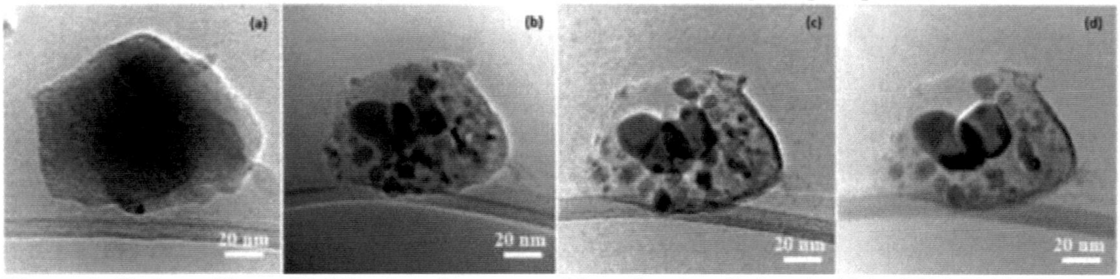

Figure 4. Zn-borosilicate (ZnBS) glass fragment subjected to irradiation (a, b) in JEM 3010 TEM at 300 kV and heating to 300° C (c, d) using a Gatan heating holder in the same TEM.

Our study aims at rank-ordering the three mostly quoted mechanisms to explain electron irradiation induced shape changes of glass nanoparticles, such as temperature, charging-fields, and radiation induced flow (RIF). Evidence has been found that generally all three mechanisms are at work and superimpose. However, in individual situations, one effect might dominate and dwarf the others. For small particles (<100 nm), temperature appears negligible and RIF explains all observations. For larger particles and where bad contact to carbon-supports is observed, beam-heating seems evident however without symptoms of true melting. In-situ TEM heating confirmed that at least up to 300°C no temperature-related re-shaping occurs at all. The side-effects observed simultaneously alongside the main glass softening include phase separation and particle precipitation. Both prove that enhanced diffusion by irradiation generated defect concentrations is a shared underlying conditions of all effects observed. The relevance of the local irradiation in Figure 4d is that due to the flat and straight geometry of the fragment edge, surface tension is eliminated as driving force. For such localised focused irradiation of an impact area within a particle, charging and field-driven ion migration/ablation, effect M2, then becomes the lead-effect as M1 is effectively symmetry-cancelled.

4. Conclusions

The observed effects are of importance for glass-surface engineering (both smoothing and - via phase separation - roughening could be achieved), but also for improving wetting-behaviour, for defect healing (low-temperature "annealing"), or changing porosity characteristics in nanoporous glasses, and ultimately also for glass patterning and grating formation.

Acknowledgements. The authors thank Iraqi Kurdistan Regional Government/Ministry of Higher Education & Scientific Research (Human Capacity Development Program), and Sheffield University, UK, for funding contributions, as well as Russell Hand for helpful discussions.

References

[1] Weber W J, et al. 1997 J. Mater. Res. **12**, 1946.
[2] Zhu X, Su J, Wu Y, Wang L, Wang Z 2014 Nanoscale **6** 1499.
[3] Mayr S G, Ashkenazy Y, Albe K, Averback R S 2003 Phys. Rev. Lett. **90** 055505.
[4] Möbus G, Ojovan M, Cook S, Tsai J, Yang G, 2010 J. Nucl. Mater. **396** 264.
[5] Jiang, N 2013 J. Phys. D: Appl., **46** 305502.
[6] Jiang, N 2016 Micron, **83** 79.
[7] Sabri M M, Hand R J, Möbus G 2015 J. Mater. Res **30** 1914.
[8] Sabri M M, Möbus G 2016 J. Nanopart. Res. **18** 1.
[9] Sabri M M, Möbus G 2017 Appl. Phys. A **123**:455.

IOP Conf. Series: Journal of Physics: Conf. Series **902** (2017) 012010 doi:10.1088/1742-6596/902/1/012010

Temperature dependence of electrical properties of electrodeposited Ni-based nanowires

R Dost, D A Allwood and B J Inkson

Dept. of Materials Science & Engineering, University of Sheffield, Sheffield, UK

E-mail: r.dost@sheffield.ac.uk

Abstract. The influence of annealing on the microstructure and the electrical properties of cylindrical nickel-based nanowires has been investigated. Nanowires of nickel of nominally 200 nm diameter and of permalloy (Py) of nominally 70 nm were fabricated by electrochemical deposition into nanoporous templates of polycarbonate and anodic alumina, respectively. Characterization was carried out on as-grown nanowires and nanowires heat treated at 650°C. Transmission electron microscopy and diffraction imaging of as-grown and annealed nanowires showed temperature-correlated grain growth of an initially nano-crystalline structure with ≤8 nm (Ni) and ≤20 nm (Py) grains towards coarser poly-crystallinity with grain sizes up to about 160 nm (Ni) and 70 nm (Py), latter being limited by the nanowire width. The electrical conductivity of individual as-grown and annealed Ni nanowires was measured in situ within a scanning electron microscope environment. At low current densities, the conductivity of annealed nanowires was estimated to have risen by a factor of about two over as-grown nanowires. We attribute this increase, at least in part, to the observed grain growth. The annealed nanowire was subsequently subjected to increasing current densities. Above 120 kA mm^{-2} the nanowire resistance started to rise. At 450 kA mm^{-2} the nanowire melted and current flow ceased.

1. Introduction

When Masuda *et al.* obtained controlled nanoporous alumina templates in 1995 [1] the fabrication of electrodeposited metallic nanowires was set to become commonplace. With time, nanowires with increasing complexity were grown, such as modulated shape [2] and composition [3, 4]. Using magnetic materials, they are still actively investigated today, mainly due to their various promising applications ranging from high-density magnetic recording [5] to high-frequency devices [6], passing by sensors [7] and biomedical applications [8]. However, the small size, fragility and particular features of cylindrical magnetic nanowires – such as high shape anisotropy, small magnetic volume, dipolar magnetic interactions between them when in array conformation, and electronic properties – present challenges to their characterization. The scientific community has been studying their structural [9], magnetic [3, 10] and electrical properties [4, 11] with great interest.

On another note, the emerging field of magnon spintronic devices promises to become a viable alternative to the stagnating complementary metal-oxide semiconductor (CMOS) based data processing technology [12, 13]. One way to help drive its miniaturization down to the nanoscale is a microwave-driven nanosized antenna that can excite high-wave-number and small-wavelength spin waves in ferromagnetic wave guides [14]. However, the choice of patterning technique in magnonic device fabrication has largely been electron-beam lithography. The more recent idea to complement magnonic structures with electrodeposited nanowires is founded on the constant strive to improve their design and efficiency. For instance, cylindrical magnetic nanowires have the potential to act as a resonant transducer that couples a uniform external microwave field to propagating spin waves. Au *et al.* have shown that the resonantly driven magnetization precession of a cylindrical transducer creates a local dynamic dipolar field that can exceed that of the incident microwave field by one order of magnitude. Coupling strength and spin wave excitation are amplified [15].

In view of controlling the functional properties of nanosized cylindrical magnetic microwave-to-spin-wave transducers, here we study the effect of heat treatment on the grain size of nickel-based cylindrical nanowires and on their electrical properties. Effects on the magnetic properties were also studied and will be published elsewhere.

2. Sample preparation

A nanoporous anodic alumina (AAO) template with about 70 nm pore diameter and 20 nm pore spacing was prepared from a 100 µm thick aluminum sheet (Goodfellow®, 99.99% pure). At first, the aluminum was polished in a 1:4 parts solution of perchloric acid and ethanol at 2–3 Adc for a few seconds, then anodized in an aqueous solution of 0.3 mol/l oxalic acid and 0.05 mol/l phosphoric acid at 33 V for 3 h. The two alumina layers were then separated with mercury(II) chloride and the barrier layer was removed in 0.5 mol/l phosphoric acid for 90 min. A VoltaLab® PGZ 402 voltammetry cell was used to perform the electrochemical deposition of Ni-based nanowires at a DC potential of -1.2 V for 45 min. Solutions for the nickel electrodeposition contained 90 g/l $NiSO_4$, 30 g/l H_3BO_3, 5 g/l L-ascorbic acid and 1 wt% gelatin, and for the electrodeposition of Ni–Fe 80/20 wt% permalloy (Py) contained 90 g/l $NiSO_4$, 13.5 g/l $FeSO_4$, 30 g/l H_3BO_3 and 5 g/l L-ascorbic acid.

Cylindrical Ni nanowires (200 nm nominal diameter) were electrochemically deposited in a single nanoporous polycarbonate template (Whatman® Nuclepore). Cylindrical Py nanowires (70 nm nominal diameter) were electrochemically deposited in the aforedescribed AAO. Figure 1a shows the nanoporous structure of an AAO template, and figure 1b depicts electrodeposited Py nanowires, which grew bottom-up about halfway into the template pores. The nanowires obtained are typically 5–9 µm long. Scanning electron microscopy (SEM) imaging was carried out on a FEI Nova™ NanoSEM 450.

Electrodeposited templates were selectively subjected to a heat treatment in a furnace under a protective N_2 atmosphere. The temperature was raised to 650°C, held for one hour, then allowed to naturally cool. We refer to *as-grown* nanowires when not heat treated and *annealed* when heat treated.

Some Ni and Py nanowires were released by dissolving their templates in chloroform and 1M NaOH solution, respectively, then drop cast from suspension onto a substrate. The nanowires dispersed randomly and came to rest either isolated or in bundles (figure 1c), depending on the interplay of surface energies and cohesive forces between individual nanowires and the substrate.

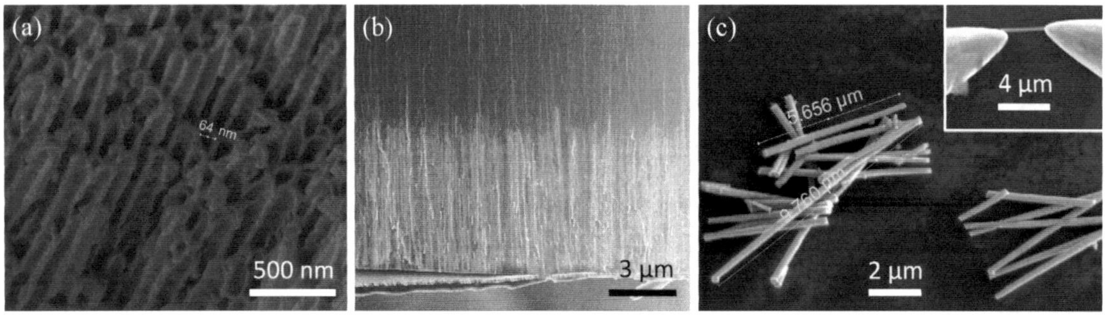

Figure 1. (a) SEM image of nanoporous AAO template. **(b)** He-ion SE image of electro-deposited AAO template. Permalloy nanowires grew bottom-up halfway inside the pores (bright contrast). **(c)** Randomly organized nickel nanowires after drop casting from chloroform suspension. Inset: Tungsten tips making electric contact with a Ni nanowire.

3. Experimental results

Transmission electron microscopy (TEM) and diffraction imaging (JEOL JEM 3010) was used to characterize the grain sizes of Ni-based nanowires prior to and after heat treatment. The as-grown nanowires are nano-crystalline. Grain sizes are about 20 nm or less for permalloy (figure 2a) and 8 nm or less for nickel (figure 3a). Heat treatment induces grain growth. After 650°C annealing, permalloy nanowires feature large grains that can fill the entire nanowire width (∅70 nm), forming a 'bamboo structure' (figure 2b) [16]. The thicker poly-crystalline nickel nanowires (∅200 nm) grow grains up to 160 nm (figure 3b). The width of the annealed nanowire remained unaffected to within ±2 nm, or 1%.

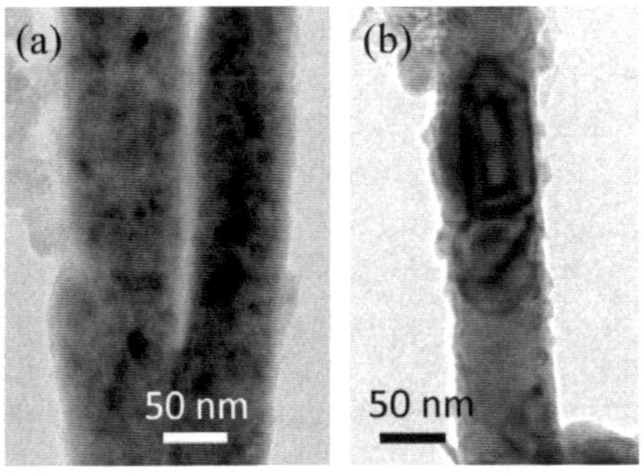

Figure 2. TEM images of ~80 nm wide Py nanowires. **(a)** Two as-grown nanowires with grain sizes ≤20 nm. **(b)** After annealing at 650°C, the nanowire grew large grains up to its full width.

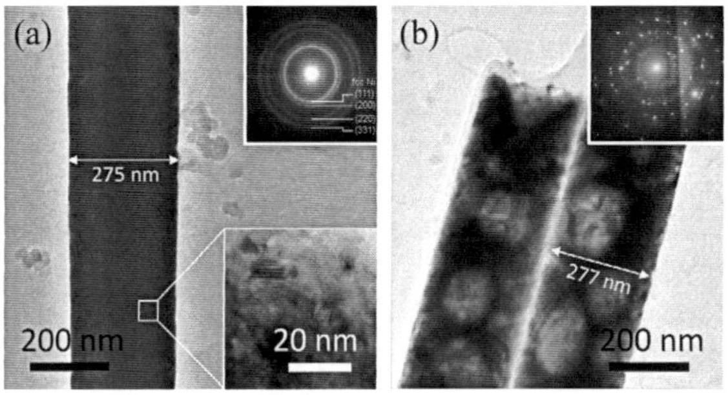

Figure 3. TEM images of ~275 nm wide Ni nanowires. **(a)** Before annealing grain sizes are ≤8 nm (see magnified area). **(b)** After annealing at 650°C, grain sizes are up to ~160 nm. Insets: Representative diffraction patterns.

The electrical conductivities of untreated and 650°C heat-treated individual Ni nanowires (275 nm diameter) were studied from current-voltage measurements. The nanowires were probed in a SEM environment by two Kleindiek MM3A-EM micromanipulators with tungsten probes (inset to figure 1c). All measurements used the same pair of probes with a tip-to-tip resistance (R_t) of about 30 Ω. To make good mechanical contact with the nanowires, the tungsten tips applied a gentle vertical. To make good electrical contact and minimize the effect of contact resistance, the actual IV measurements preceded one or more 'burn-in' bias sweeps where voltage was raised stepwise (up to ±5 V) until total resistance (R) reached a minimum. This process breaks down any oxide layer that may have been present on the nanowire surface [17] and yielded a repeatable total resistance to within ±2% for the same kind of nanowires and same probe spacing ($R_{nw} \cong$ const). By implication, the uncertainty in R_c had to also be approximately equal to ±0.02 R – for any nanowire – albeit of unknown magnitude.

Figure 4a shows the IV curve of an as-grown nanowire at low forward-reverse biasing at 2.6 μm probe spacing (L_1) and figure 4b of an annealed nanowire at maximum forward biasing at 2.9 μm probe spacing (L_2).The as-grown nanowire exhibits a linear $I(V)$ characteristic within the applied ±0.1 V sweep, corresponding to current densities (J) of 0–43 kA mm^{-2}, with a total resistance R_1 = 38.7 Ω. The 650°C annealed nanowire exhibited four distinct regimes during the high voltage sweep (see ①–④ in figure 4b).

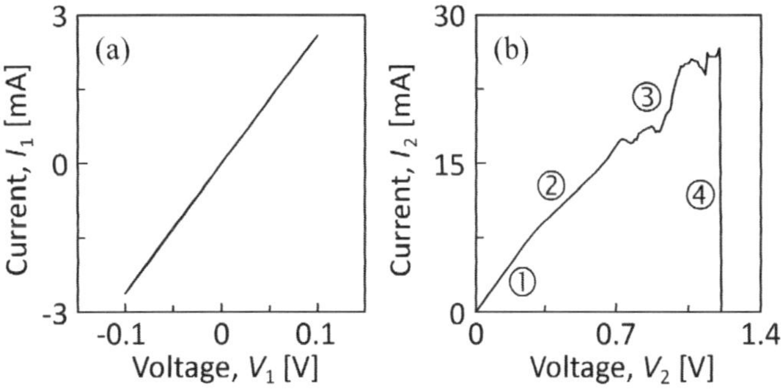

Figure 4. IV curves of **(a)** an individual as-grown Ni nanowire and **(b)** of an individual Ni nanowire annealed at 650°C.

At low biasing, corresponding to J = 0–120 kA mm^{-2}, total resistance R_2 was 34.9 Ω and ohmic in nature. This change corresponds to a 9.9% drop in total resistance for the annealed nanowire and cannot be fully ascribed to uncertainty in contact resistance alone, estimated at ±0.7 Ω for either measurement. At increased biasing, equal to J = 120–240 kA mm^{-2}, R_2 rose to 51.7 Ω. Above 240 kA mm^{-2}, R_2 fluctuated erratically until current flow abruptly ceased at 450 kA mm^{-2}. As observed in situ during SEM imaging, the nanowire melted from excessive Joule heating.

The change in conductivity (σ) between as-grown and annealed nanowires was estimated for an unknown but constant voltage drop (V_c) across R_c by evaluating both IV curves in the low biasing regime at identical currents of 2.58 mA (arbitrary). Given the same cross-sectional areas, the conductivity ratio of annealed (σ_{nw2}) to as-grown nanowire (σ_{nw1}) is $\frac{\sigma_{nw2}}{\sigma_{nw1}} = \frac{L_2}{L_1}\left(\frac{R_1}{R_2} - \frac{R_c}{R_2}\right)\left(1 - \frac{R_c}{R_2}\right)^{-1}$. Assuming $R_c \sim R_t$, the 9.9% drop in total resistance corresponds to an increase in the annealed nanowire conductivity by +99%. Should R_c approach 0.9 R_2 then σ_{nw2} approached the Ni bulk conductivity of 1.43×10^7 S m^{-1} [18], corresponding to an an increase in conductivity by +136%.

We conclude, the estimated increase in nanowire conductivity must be attributed, at least in part, to the observed grain growth with heat treatment. Larger grains mean fewer grain boundaries – sites of electron scattering, which contribute to electrical resistance. Magneto-optical Kerr effect magnetometry showed that the Ni and Py nanowires exhibit magnetic properties. The data will be detailed elsewhere.

4. Conclusion

Here we show that the grain structure of magnetic cylindrical nanowires can be altered through heat treatment, which causes significant grain growth, tending towards a bamboo-microstructure. This in turn reduces grain boundary surface area with an approximately two-fold increase in electrical conductivity for the 650°C annealed nickel nanowires. Current densities above about 120 kA mm^{-2} lead to Joule heating and degrade conductivity until structural failure.

Acknowledgement
This work was supported by the Engineering and Physical Sciences Research Council [grant number EP/L020696/1].

References

[1] Masuda H and Fukuda K 1995 *Science* **268** pp 1466–8
[2] Lee W, Ji R, Gösele U and Nielsch K 2006 *Nat. Mater.* **5** pp 741–7
[3] Moura K O *et al.* 2016 *Sci. Rep.* **6** 28364
[4] Elawayeb M, Peng Y, Briston K J and Inkson B J 2012 *J. Appl. Phys.* **111** 034306
[5] Irshad M I, Ahmad F and Mohamed N M 2012 *AIP Conf. Proc.* (Malaysia. Melville: AIP) p 625
[6] Frake J C, Kano S, Ciccarelli C, Griffiths J, Sakamoto M, Teranishi T, Majima Y, Smith C G and Buitelaar M R 2015 *Sci. Rep.* **5** pp 1–6
[7] Chena X, Wonga C K Y, Yuan C A and Zhanga G 2013 *Sensor Actuat.* B: Chem. **177** pp 178–95
[8] Patolsky F, Zheng G and Lieber C M 2006 *Nanomedicine* **1** pp 51–65
[9] Maaz K *et al.* 2010 *Nanoscale Res. Lett.* **5** pp 1111–17
[10] Encinas-Oropesa A, Demand M and Piraux L 2001 *Phys. Rev.* B **63** 104415
[11] Wu J, Yin B, Wu F, Myung Y and Banerjee P 2014 *Appl. Phys. Lett.* **105** 183506
[12] Kruglyak V V, Demokritov S O and Grundler D 2010 *J. Phys.* D: Appl. Phys. **43** 264001
[13] Lenk B, Ulrichs H, Garbs F and Münzenberg M 2011 *Phys. Rep.* **507** pp 107–36
[14] Ciubotaru F 2012 Thesis: *Spin-wave excitation by nano-sized antennas* (University of Kaiserslautern)
[15] Au Y, Ahmad E, Dmytriiev O, Dvornik M, Davison T and Kruglyak V V 2012 *Appl. Phys. Lett.* **100** 182404
[16] Hummelgård M *et al.* 2010 *Nanotechnology* **21** 165704
[17] Krishnan S, Stefanakos E and Bhansali S 2007 Thin Solid Films **516** pp 2244–50
[18] Laubitz M J, Matsumura T and Kelly P J 1976 *Can. J. Phys.* **54** pp 92–102

Measurement of Off-Axial Aberration

Hidetaka Sawada[1,2,3], and Angus I. Kirkland[1,2]

[1]ePSIC: Electron Physical Sciences Imaging Centre, Diamond Light Source Ltd, Didcot, Oxford, OX11 0DE, UK
[2]Department of Materials, University of Oxford, Parks Road, Oxford OX1 3PH, UK
[3]JEOL UK Ltd., Silver Court Watchmead Welwyn Garden City, Herts.,AL7 1LT, UK

E-mail: hide.sawada@materials.ox.ac.uk

Abstract. High-resolution observation of a large field of view with a high quality detector is important for single particle analysis or in-situ experiments. Consequently, evaluation of aberrations that depend on position in the field of view, (off-axial aberrations) is required. We report the experimental measurement of off-axial aberrations including higher order components in several optical conditions together with the corresponding off-axial phase maps.

1. Introduction

Owing to the development of a high quality cameras with a large number of pixels, high resolution observations in both materials science and life science over a large field of view is becoming important in transmission electron microscopy (TEM). Atomic-resolution observation of two-dimensional materials and in-situ studies in the TEM require optimization of the electron-optical conditions over a large field of view, for example to understand point defects, dislocations and distortions of the lattice [1]. High-resolution image recording for single particle analysis with wide field of view cameras is also necessary to image a suitably large number of particles for reconstruction. However, when recording high-resolution TEM data with a large field of view, off-axial aberrations increase in the peripheral area, corrupting the data. Consequently, high-quality imaging is guaranteed only at the center axis of the image. Evaluation of off-axial aberrations is hence required to optimize the optical conditions for wide field of view imaging.

Measurement of off-axial aberrations using diffractogram from different positions in the experimental image field of view has been previously reported [2]. In this study, we define the off-axial aberrations, and extend the measurement of these to defocus and two-fold astigmatism up to 4th rank. Features arising from of off-axial aberrations are discussed using the simulated patterns of diffractograms and compared to experimental measurements using an off-axial aberration phase map.

2. Measurement of off-axial aberrations

Aberrations including a spatial vector, $\mathbf{r} = (x, y)$ are defined as off-axial aberrations (Fig.1(b)), whereas aberrations that are described by only a complex angle, $\boldsymbol{\omega} = \boldsymbol{\alpha} \cdot \mathbf{exp(i\theta)}$ are defined as axial aberrations (Fig.1(a)), where α is the amplitude of the angle and θ is the azimuth. The total wave aberration function, $\boldsymbol{W}_{total}$ is described by summation of the axial and off-axial aberrations, as;

$$\boldsymbol{W}_{total} = \boldsymbol{W}_{axial} + \boldsymbol{W}_{off\ axial} \quad (1)$$

where $\boldsymbol{W}_{axial}$ is the axial wave aberration function and $\boldsymbol{W}_{off\ axial}$ is the off-axial wave aberration function.
The axial wave aberration function can be rewritten as;

$$\boldsymbol{W}_{axial} = \sum_{m,n} \frac{1}{m+n} \mathrm{Re}(\boldsymbol{C}_{m,n} \overline{\boldsymbol{\omega}}^m \boldsymbol{\omega}^n) \quad (2)$$

where $\overline{\boldsymbol{\omega}} = \alpha \cdot \exp(-i\theta)$ is the complex conjugate of $\boldsymbol{\omega}$.
The off-axial aberration wave function is given by;

$$\boldsymbol{W}_{off\ axial} = \sum_{m,n,s,t} \frac{1}{m+n} \mathrm{Re}\left(\boldsymbol{K}_{m,n,s,t} \overline{\boldsymbol{\omega}}^m \boldsymbol{\omega}^n \overline{\boldsymbol{r}}^s \boldsymbol{r}^t\right) \quad (3)$$

where $\mathbf{r} = |r| \cdot \exp(i\theta_r)$ is a complex position, $|r|$ is amplitude of the complex position and θ_r is the azimuth of the complex position. In order to extend measurements to higher orders of off-axial aberration coefficients, a dimensionless parameter can be conveniently introduced to give the same unit for all coefficients. Hence, a dimensionless angle, $\boldsymbol{\gamma}$ is introduced as a parameter for the spatial position and consequently the off axial wave aberration function, $\boldsymbol{W}_{off\ axial}$ can be rewritten as;

$$\boldsymbol{W}_{off\ axial} = \sum_{m,n,s,t} \frac{1}{m+n} \mathrm{Re}\left(\boldsymbol{K}_{m,n,s,t(r0)} \overline{\boldsymbol{\omega}}^m \boldsymbol{\omega}^n \overline{\boldsymbol{\gamma}}^s \boldsymbol{\gamma}^t\right) \quad (4)$$

where the complex dimensionless position $\boldsymbol{\gamma}$ at r_0; r_0 is a standard length; $\boldsymbol{\gamma} = |\gamma| \cdot \exp(i\theta_\gamma)$ with $|\gamma| = |r|/|r_0|$, which is the amplitude of the complex dimensionless position and $\theta_\gamma = \theta_r$ is the azimuth of the complex position. The geometrical order, N of the axial aberrations is $\mathrm{N} = \mathrm{m} + \mathrm{n} - 1$, and R is a rank including the off-axial aberrations with $\mathrm{R} = \mathrm{N} + \mathrm{s} + \mathrm{t}$ [3]. The off-axial aberration coefficient of $\boldsymbol{K}_{m,n}$ at r_0 for an axial aberration $\boldsymbol{C}_{(m+n)}$ can be described by summation of each rank of the off-axial aberrations as,

$$\boldsymbol{K}_{m,n} = \sum_{s,t} \boldsymbol{K}_{m,n,s,t(r0)} \overline{\boldsymbol{\gamma}}^s \boldsymbol{\gamma}^t \quad (5)$$

Hence, each coefficient (vector) of the off-axial aberrations for $(\mathrm{m}, \mathrm{n}, \mathrm{s}, \mathrm{t})$ at r_0 is the term corresponding to complex position of $\overline{\boldsymbol{\gamma}}^s \boldsymbol{\gamma}^t$ defined as $\boldsymbol{K}_{m,n,s,t(r0)}$. For the data reported, $\boldsymbol{\gamma}$ is defined at r_0=100 nm whence an off axial aberration coefficient corresponds to a value of the axial aberration 100 nm away from the centre of the axial position with a unit of length in this case.

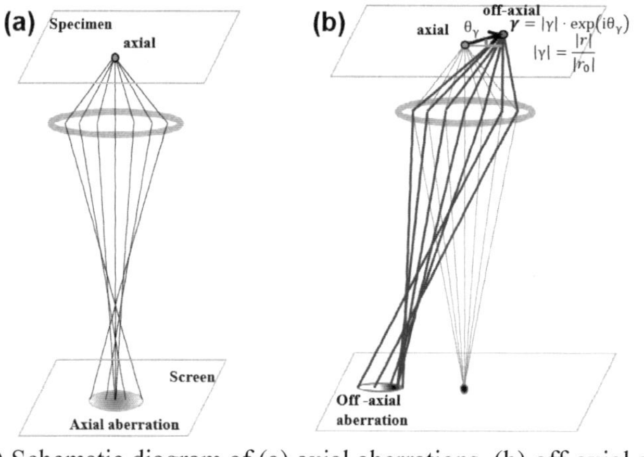

Figure 1. (a) Schematic diagram of (a) axial aberrations. (b) off-axial aberrations.

Table 1: Off-axial aberration coefficients. Using corresponding axial aberration notations, $\mathbf{O}_2$ denotes defocus, $\mathbf{A}_2$ denotes two-fold astigmatism, $\mathbf{P}_3$ denotes axial coma, and $\mathbf{A}_3$ denotes three-fold astigmatism.

	Off-axial aberration $\boldsymbol{K}_{m,n,s,t(r0)}$			
Axial aberration	Off-axial aberration	2nd rank	3rd rank	4th rank
$\mathbf{O}_2$	$\mathrm{Re}[\boldsymbol{O}_{2\,\overline{\gamma}^s\gamma^t(r0)}]$	$\mathbf{O}_{2\,\gamma(r0)}$	$\mathbf{O}_{2\,\gamma\overline{\gamma}(r0)}$, $\mathbf{O}_{2\,\gamma^2(r0)}$	$\mathbf{O}_{2\,\gamma^2\overline{\gamma}(r0)}$, $\mathbf{O}_{2\,\gamma^3(r0)}$
$\mathbf{A}_2$	$\boldsymbol{A}_{2\,\overline{\gamma}^s\gamma^t(r0)}$	$\mathbf{A}_{2\,\gamma(r0)}$, $\mathbf{A}_{2\,\overline{\gamma}(r0)}$	$\mathbf{A}_{2\,\gamma\overline{\gamma}(r0)}$, $\mathbf{A}_{2\,\gamma^2(r0)}$, $\mathbf{A}_{2\,\overline{\gamma}^2(r0)}$	
$\mathbf{P}_3$	$\boldsymbol{P}_{3\,\overline{\gamma}^s\gamma^t(r0)}$		$\mathbf{P}_{3\,\gamma(r0)}$, $\mathbf{P}_{3\,\overline{\gamma}(r0)}$	
$\mathbf{A}_3$	$\boldsymbol{A}_{3\,\overline{\gamma}^s\gamma^t(r0)}$		$\mathbf{A}_{3\,\gamma(r0)}$, $\mathbf{A}_{3\,\overline{\gamma}(r0)}$	

Table 2: Coefficients of off-axial defocus and off-axial two-fold astigmatism

Symmetry → / Rank ↓	0	1	2	3	0	1	2	3	4
	Off-axial defocus				Off-axial two-fold astigmatism				
2		$\mathbf{O}_{2\,\gamma(r0)}$				$\mathbf{A}_{2\,\gamma(r0)}$		$\mathbf{A}_{2\,\bar{\gamma}(r0)}$	
3	$\mathbf{O}_{2\,\gamma\bar{\gamma}(r0)}$		$\mathbf{O}_{2\,\gamma^2(r0)}$		$\mathbf{A}_{2\,\gamma^2(r0)}$		$\mathbf{A}_{2\,\gamma\bar{\gamma}(r0)}$		$\mathbf{A}_{2\,\bar{\gamma}^2(r0)}$
4		$\mathbf{O}_{2\,\gamma^2\bar{\gamma}(r0)}$		$\mathbf{O}_{2\,\gamma^3(r0)}$					

Diffractograms at different positions in a field of view were simulated using second rank off-axial defocus and off-axial two-fold astigmatism (Fig.2). Figure 2(a) shows diffractograms of segmented areas from the full field of view with off-axial defocus, $\mathbf{O}_{2\,\gamma(100nm)}$ = (100nm,0) and axial defocus, $\mathbf{O}_2$ = –100 nm. The defocus with position in the field of view is determined by the off-axial defocus and the defocus becomes close to zero at the right hand side. Figure 2(b) shows diffractograms from segmented areas with $\mathbf{O}_{2\,\gamma(100nm)}$ =(0,100nm), where the defocus at the lower part is close to zero. The position dependence of defoci due to the off-axial defocus $\mathbf{O}_{2\,\gamma(100nm)}$ = (0,100nm) shows a 90° rotation relationship to the off-axial defocus $\mathbf{O}_{2\,\gamma(100nm)}$ =(100nm,0). Figs. 2(c) and 2(d) show diffractograms from segmented areas with off-axial two-fold astigmatism, $\mathbf{A}_{2\gamma}$ and $\mathbf{A}_{2\bar{\gamma}}$, respectively. The two-fold astigmatism becomes larger toward the edge of the field of view, and the azimuth of the two-fold astigmatism changes depending on position. A map of $\mathbf{A}_{2\bar{\gamma}}$ shows an overall three-fold symmetry, since the symmetry of the pattern in the field of view is |(m-n) + (s-t)| in the case of off-axial defocus and off-axial two-fold astigmatism. Table 3 lists examples of measured coefficients for the off-axial aberrations using experimental diffractograms at a 200kV using JEM-ARM300F.

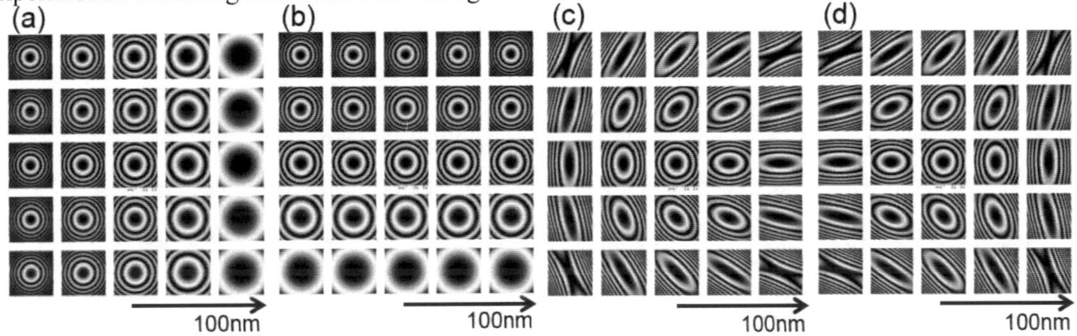

Figure 2. Diffractograms from segmented areas with a width of 200 nm (a)$\mathbf{O}_{2\,\gamma(100nm)}$ = (100nm,0), (b) $\mathbf{O}_{2\,\gamma(100nm)}$ = (0,100nm), (c) $\mathbf{A}_{2\,\gamma(100nm)}$ = (100nm, 0) (d) $\mathbf{A}_{2\,\bar{\gamma}(100nm)}$ = (0, –100nm).

Table 3. Measured values of off axial aberration coefficients. (r_0=100 [nm])

Rank	Off-axial aberration	Amplitude	Azimuth	Rank	Off-axial aberration	Amplitude	Azimuth
2	$O_{2\,\gamma(r0)}$	8.620 nm	–102.00°	2	$A_{2\,\gamma(r0)}$	1.9067 nm	–156.65°
3	$O_{2\,\gamma\bar{\gamma}(r0)}$	–0.3148 nm		2	$A_{2\,\bar{\gamma}(r0)}$	1.1012 nm	–18.94°
3	$O_{2\,\gamma^2(r0)}$	0.4245 nm	–24.02°	3	$A_{2\,\gamma^2(r0)}$	0.7866 nm	165.90°
4	$O_{2\,\gamma^2\bar{\gamma}(r0)}$	0.1341 nm	177.86°	3	$A_{2\,\gamma\bar{\gamma}(r0)}$	0.5034 nm	12.55°
4	$O_{2\,\gamma^3(r0)}$	0.0264 nm	–67.50°	3	$A_{2\,\bar{\gamma}^2(r0)}$	0.5401 nm	–92.60°
3	$P_{3\,\gamma(r0)}$	407.66 nm	–30.10°	3	$A_{3\,\gamma(r0)}$	5.963 nm	61.76°
3	$P_{3\,\bar{\gamma}(r0)}$	50.907 nm	–77.95°	3	$A_{3\,\bar{\gamma}(r0)}$	13.317 nm	154.36°

3. Experimental measurement of off-axial aberrations

Figs. 3(d-f) show off-axial phase maps from experimental measurements. Figs.3(a-c) were recorded by changing the excitation of the transfer lens (TL) between two multipoles in the TEM corrector. Fig.3(a)

was recorded using a weaker excitation of the TL, and Fig.3(c) with a stronger excitation. The TL excitation used to record Fig.3(b) is intermediate between that used to record Figs.3(a) and 3(c). In general, specimen or local bending introduces a defocus change across the field of view, artificially resulting in second rank off-axial defocus. To distinguish genuine electron-optical off-axial aberrations from this specimen effect, the second rank off-axial defocus was set to zero in the phase maps shown in Figs.3(d-f). From these measurements, the value of a $A_{2\bar{\gamma}}$, which has three-fold symmetry, shows a significant difference (Table 4). The off-axial aberrations cause a change of the two-fold astigmatism at the periphery of the field of view demonstrating that optimization of $A_{2\bar{\gamma}}$ is important for a double hexapole TEM corrector.

Table 4: Measured values of off axial aberrations. (r_0=100 [nm])

	Setting 1(Figs.3(a) and 3(d))		Setting 2(Figs.3(b) and 3(e))		Setting 3(Figs.3(c) and 3(f))	
Off-axial aberration	Amplitude	Azimuth	Amplitude	Azimuth	Amplitude	Azimuth
$A_{2\gamma(r_0)}$	1.81 nm	142.59°	1.83 nm	−125.38°	2.67 nm	−110.61°
$A_{2\bar{\gamma}(r_0)}$	**4.38** nm	72.71°	**1.04** nm	101.96°	**2.70** nm	−131.97°

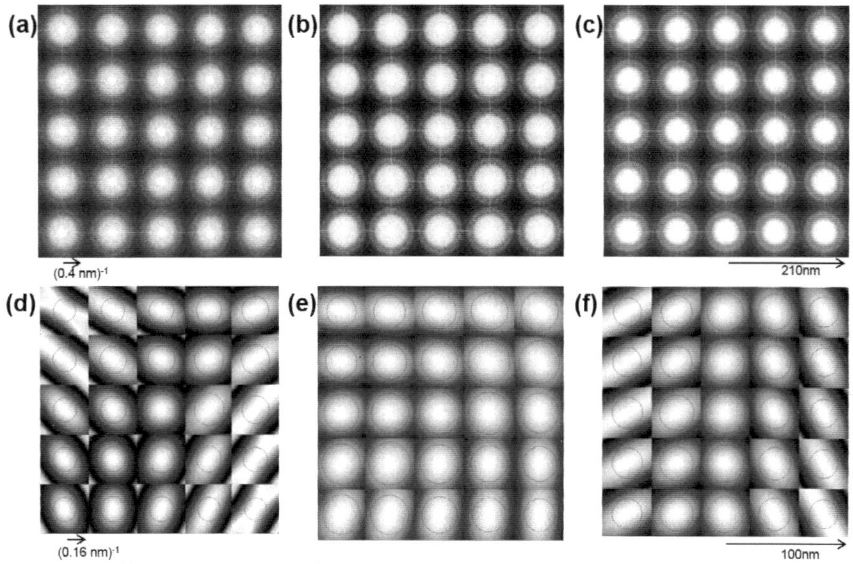

Figure 3. (a-c) 5x5 diffractogram maps from segmented areas. (d-f) Off-Axial Phase Map for different TL excitations. An offset, O_2= −5 nm is used to visualize the off-axial two-fold astigmatism in the phase maps. Red circles denote the $\pi/4$ phase limit within each local area.

4. Conclusion

Measurement of off-axial aberrations including those of higher rank was reported using diffractograms calculated from different positions in a wide field of view. Experimental measurements have been reported for several optical settings. For the measurements reported off-axial two-fold astigmatism, $A_{2\bar{\gamma}}$, with a three-fold symmetry across the field of view is clearly evident in images recorded using a double-hexapole TEM corrector.

References

[1] Warner J H, Margine E R, Mukai M, Robertson A W, Giustino F and Kirkland A I 2012 *Science* **337** 209-212.

[2] Maßmann I, Uhlemann S, Müller H, Hartel P, Zach J, Haider M, Taniguchi Y, Hoyle D and Herring E 2011 *Microsc. Microanal.* **17** Suppl. 1270.

[3] Hawkes P W and Kasper E 1989 Principles of Electron Optics. Academic Press. London. UK.

IOP Conf. Series: Journal of Physics: Conf. Series **902** (2017) 012012 doi:10.1088/1742-6596/902/1/012012

Accurate Evaluation of Aberration for Probe-Forming System and Influence of Aberration on High-Resolution STEM image

Hidetaka Sawada[1,2,3], Fumio Hosokawa[4] and Angus I. Kirkland[1,2]

[1]ePSIC: Electron Physical Sciences Imaging Centre, Diamond Light Source Ltd, Didcot, Oxford, OX11 0DE, UK
[2]Department of Materials, University of Oxford, Parks Road, Oxford OX1 3PH, UK
[3]JEOL UK Ltd., Silver Court Watchmead Welwyn Garden City, Herts., AL7 1LT, UK
[4]BioNet Laboratory Inc., 2-3-28 Nishiki-chou, Tachikawa, Tokyo, 190-0022 Japan

E-mail: hide.sawada@materials.ox.ac.uk

Abstract. The influence of residual three-fold astigmatism on high-resolution STEM imaging of graphene has been investigated. The aberration was measured by the segmented Ronchigram auto-correlation function method using the graphene lattice. In the presence of residual three-fold astigmatism the graphene lattice showed a lower three-fold rotational symmetry where three C atoms in the six membered ring had a higher intensity.

1. Introduction

Aberration corrected instruments are commonly used to obtain high-resolution imaging in both STEM and TEM [1,2], and single atoms of light elements including boron, carbon, nitrogen, and oxygen can now be imaged in two-dimensional materials. For visualization of single atoms in two-dimensional materials, residual aberrations affect the image quality in STEM by altering the probe shape. Automated measurement and adjustment of the aberrations in the probe-forming system can be achieved using pairs of over and under focus images of nm sized metal particles supported on a carbon film by deconvolution of probe shapes using a Gaussian focus image [1]. Alternatively, the Ronchigram can be used to measure aberrations with beam induced image shifts [2] and segmentation of Ronchigrams recorded from amorphous films [3] or lattices [4, 5]. However, even after auto-adjustment [1-3], residual aberration can be introduced by the hysteresis in the objective lens, specimen charging and other factors. During high-resolution image observation it is possible to manually correct first-order defocus and two-fold astigmatism to optimize these aberration. Second-order axial coma can also generally be corrected manually by observing the Rongchigram. However, second-order three-fold stigmatism is not easy to correct manually by observing a high-resolution image or a Ronchigram. Nonetheless, three-fold astigmatism can introduce artifacts in high-resolution images, potentially leading to incorrect interpretation for quantitative analysis. We have developed a method to measure this aberration using a two-dimensional lattice [6]. Importantly using this method, a high-resolution image of a two-dimensional material and the aberration measurement can be performed with the same probe-forming condition and the same specimen. It is useful to understand the influence of this aberration directly on high-resolution STEM images and hence we have evaluated its effects using STEM images of monolayer graphene.

2. Aberration measurement and high-resolution observation of graphene

Ronchigrams from an amorphous film (Fig.1(a)) and from monolayer graphene ((Fig.1(b)) were recorded using a JEM-ARM300F equipped with a STEM spherical aberration corrector at an accelerating voltage of 80 kV. The Ronchigram of graphene was divided into local angular areas and the lateral magnification change for each segment was measured using the auto-correlation function (Fig 1(c)). Aberration coefficients were measured from the auto-correlation patterns and the values of the aberration coefficients up to 3rd order are listed in Table 1. Figure 1(d) shows a high-resolution low-angle-annular-dark-field (LAADF) STEM image of a monolayer graphene sample. The C-C spacing at 142 pm are clearly resolved and the intensity of all carbon atoms in the 6-membered rings are equivalent, as shown in the intensity line profile in Fig. 1(e).

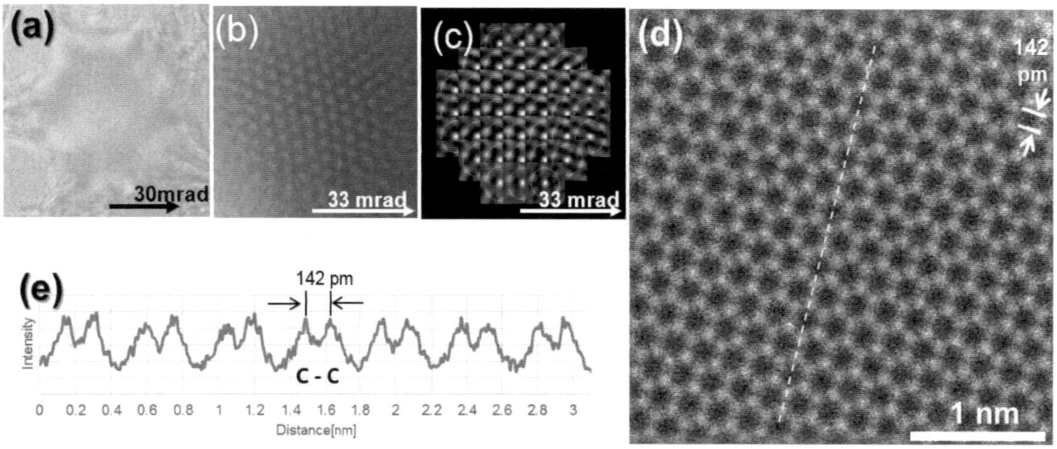

Figure 1. (a) Ronchigrams obtained from an amorphous film at 80kV with a camera length of 80 cm. (b) Experimental Ronchigram recorded from monolayer graphene. (c) Local area auto-correlation patterns from an array of 37 sub-areas. (d) LAADF STEM image of graphene at an accelerating voltage of 80kV with a convergence semi-angle of 25 mrad with 23 pA. 30 frames of 512 x 512 pixels were averaged to produce the image in (e) with a dwell time of 4 μs/pixel for each frame. (e) Intensity line profile from the dotted line in Fig.1(d) with 10 pixel width.

Table 1. Aberration coefficients up to third order from the measurements shown in Figure 1(c). Values in brackets correspond to errors in the measurement values.

Aberration	Amplitude	Azimuth
O2: Defocus	–51.852 [nm] (0.11 [nm])	
A2: Two-fold stigmatism	0.276 [nm] (0.09 [nm])	84.34° (7.57 %)
P3: Axial Coma	17.598 [nm] (3.46 [nm])	–144.39° (3.67 %)
A3: Three-fold astigmatism	16.436 [nm] (2.42 [nm])	–0.19° (2.15 %)
A4: Four-fold astigmatism	0.173 [μm] (0.08 [μm])	29.66° (7.33 %)
Q4: Star aberration	0.654 [μm] (0.16 [μm])	53.16° (4.57 %)
O4: 3rd order spherical aberration	0.833 [μm] (0.15 [μm])	

3. High-resolution STEM image with a three-fold astigmatism

Figures 2(a) and 2(d) show Ronchigrams recorded at 80 kV with a camera length of 80 cm from monolayer graphene with a focused STEM probe in which a three-fold astigmatism was intentionally introduced. The Ronchigram was segmented into 21 local angular areas and auto-correlation patterns were calculated from each area, shown in Figures 2(b) and 2(e). Table 2 lists the aberration coefficients up to third order calculated from the oval fits, showing a large value of the three-fold astigmatism. Figures 2(c) and 2(f) show the calculated phase using the measured aberrations in Table 2. In the phase map the three fold symmetry due to A3 is dominant and the polarity of the three-fold astigmatism is opposite in Figs.2(c) and 2(f).

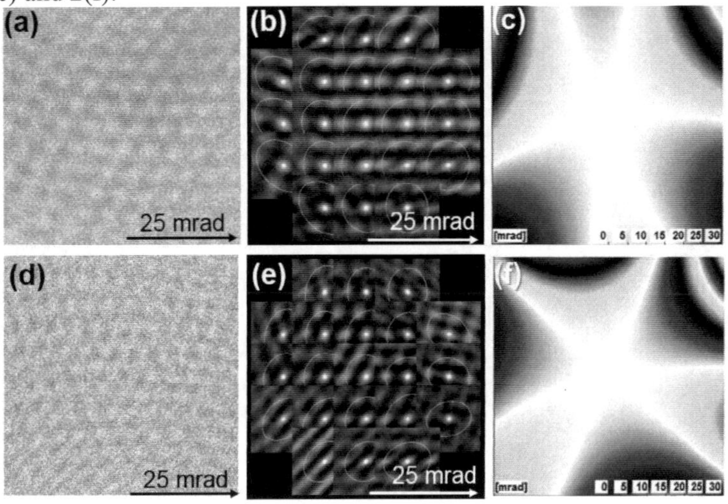

Figure 2. (a) and (d) Experimental Ronchigrams recorded from monolayer graphene with three-fold astigmatism. The Ronchigrams are shown up to 25 mrad. (b) and (e) Local area auto-correlation functions patterns calculated from the Ronchigrams in (a) and (d) with corresponding oval fits. (c) and (f) Calculated phases using the measured aberrations from (b) and (e), respectively (Table 2). White to black corresponds to π. Blue and light green show the opposite sign of the phase. Defocus and two-fold astigmatism are set to zero for the phase simulation, since first-order aberrations were adjusted during high-resolution STEM observation.

Table 2. Measured aberration coefficients from Fig. 2(b) and in Fig. 2(e).

Aberration	Amplitude in Fig.2(b)	Azimuth	Amplitude in Fig.2(e)	Azimuth
O2	−52.039 [nm] (0.47 [nm])		−59.455 [nm] (0.52 [nm])	
A2	1.381 [nm] (0.50 [nm])	−43.47° (5.41 %)	4.335 [nm] (0.56 [nm])	−47.46° (1.41 %)
P3	80.055 [nm] (11.48 [nm])	108.91° (2.45 %)	80.113 [nm] (16.73 [nm])	114.56° (2.08 %)
A3	**113.655** [nm] (11.63 [nm])	**38.32°** (1.29 %)	**191.105** [nm] (11.19 [nm])	**−17.63°** (0.70 %)
A4:	0.141 [μm] (0.40 [μm])	−35.25° (50.86 %)	1.180 [μm] (0.55 [μm])	− 35.25° (1.32 %)
Q4	1.419 [μm] (0.76 [μm])	−4.02° (8.02 %)	2.407 [μm] (0.73 [μm])	− 4.02° (5.94 %)
O4	1.412 [μm] (0.59 [μm])		−0.985 [μm] (0.76 [μm])	

STEM images of monolayer graphene in Figs.3(a) and (e) were recorded with a three-fold astigmatism (Table 2). Three atomic positions show brighter intensity within the 6-member C rings (Fig.3(a)); the carbon at position "A" is brighter than that at position "B", although all atomic position corresponds to a single carbon atom (Fig. 3(b)). Figs. 3(c) and (d) show a simulated STEM LAADF image of monolayer graphene and the corresponding simulated probe with A3=113.655 nm at 38.32° [7]. The atomic position at "A" is also brighter than at "B" in these simulations. However, the STEM image for the opposite polarity of a A3 (191.105 nm at -17.63°) (Fig.3(e)) shows that the atomic position "B" is brighter than that at "A" in both the simulated and experimental images (Figs.3(f) and (g)). Importantly, for both cases the 6-fold rotational symmetry of graphene is lowered to a 3-fold rotational symmetry. These results clearly demonstrate that aberration introduce significant artefacts in quantitative interpretation for STEM imaging.

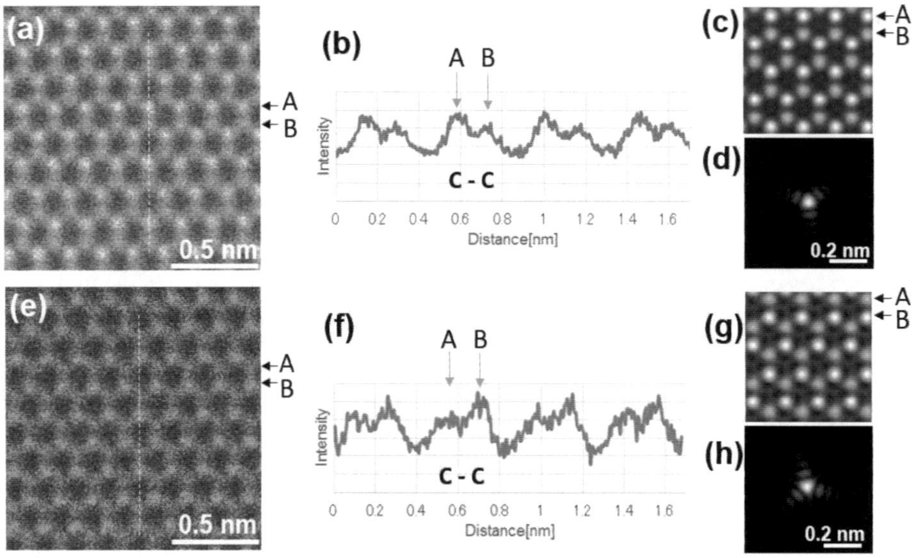

Figure 3. (a) Experimental LAADF STEM image of graphene at an accelerating voltage of 80kV with a convergence semi-angle of 25mrad and a beam current of 23 pA with an A3 value corresponding to Fig.2(a). Ten frames were averaged with a dwell time of 20 μs/pixel for each frame. (b) Intensity line profile from the white dotted line with a 10 pixel width from Fig.3(a). (c) Simulated LAADF image with A3=113.655 nm at 38.32°. (d) Simulated probe for the same value of A3 as in (c). (e) Experimental LAADF STEM image of graphene with an A3 value corresponding to Fig.2(d). Ten frames were averaged with a dwell time of 20 μs/pixel for each frame. (f) Intensity line profile from the white dotted line with a 10 pixel width from Fig.3(e). (g) Simulated LAADF image with A3=191.105 nm at -17.63°. (f) Simulated probe for the same value of A3 as in (g)

4. Conclusion

Three-fold astigmatism was accurately measured using the segmented Ronchigram auto-correlation function patterns recorded from a graphene lattice. LAADF images of a monolayer graphene in the presence of this aberration showed a lower three-fold rotational symmetry. Moreover, when the azimuth of a residual three-fold aberration was rotated, the positon of the atoms having stronger intensity were changed.

.

References

[1] Haider M., Uhlemann S, and Zach J 2000 *Ultramicroscopy* **81** 163-175.
[2] Dellby N, Krivanek, O L, Nellist P D, Batson, P E and Lupini A. R 2001 *J. Electron Microsc.*, **50**, 177-185.
[3] Sawada H, Sannomiya T, Hosokawa F, Nakamichi T, Kaneyama, Tomita T, Kondo Y, Tanaka T, Oshima Y, Tanishiro Y, and Takayanagi K. 2008 *Ultramicroscopy* **108** 1467-1475.
[4] Lupini A R and Pennycook S J 2008 *J. Electron Microsc.*, **57** 195-201.
[5] Yamazaki T, Kotaka Y, Kikuchi Y and Watanabe, K 2006 *Ultramicroscopy,* **106**, 153-163.
[6] Sawada H, Allen S.C, Wang S, Warner J H, and Kirkland A I 2017 *Ultramicroscopy,* accepted.
[7] elbis:Electron-Beam Image Simulator for Aberration Corrected TEM/ STEM and Diffractions (http://www.bio-net.co.jp/SP-elbis_e/index.html)

StatSTEM: An efficient program for accurate and precise model-based quantification of atomic resolution electron microscopy images

A De Backer[1], K H W van den Bos[1], W Van den Broek[2], J Sijbers[3], and S Van Aert [1]

[1] Electron Microscopy for Materials Science (EMAT), University of Antwerp, Groenenborgerlaan 171, 2020 Antwerp, Belgium
[2] Ag Strukturforschung/Elektronenmikroskopie, Institut für Physik, Humboldt-Universität zu Berlin, Newtonstraße, 12489 Berlin, Germany
[3] imec-Vision Lab, University of Antwerp, Universiteitsplein 1, 2610 Antwerp, Belgium

E-mail: annick.debacker@uantwerpen.be

Abstract. An efficient model-based estimation algorithm is introduced in order to quantify the atomic column positions and intensities from atomic resolution (scanning) transmission electron microscopy ((S)TEM) images. This algorithm uses the least squares estimator on image segments containing individual columns fully accounting for the overlap between neighbouring columns, enabling the analysis of a large field of view. For this algorithm, the accuracy and precision with which measurements for the atomic column positions and scattering cross-sections from annular dark field (ADF) STEM images can be estimated, is investigated. The highest attainable precision is reached even for low dose images. Furthermore, advantages of the model-based approach taking into account overlap between neighbouring columns are highlighted. To provide end-users this well-established quantification method, a user friendly program, StatSTEM, is developed which is freely available under a GNU public license.

1. Introduction

Nowadays, the field of nanotechnology requires more and more quantitative characterisations of nanomaterials. This includes for example locating atomic column positions with picometre precision [1], counting the number of atoms with single atom sensitivity [2–4] and precise chemical composition determination [5, 6]. In order to facilitate these needs, we developed a user-friendly software package called StatSTEM [7]. This program allows to apply advanced image quantification methods , such as atom-counting, measuring atomic column positions, and strain-mapping, to atomic resolution electron microscopy images by using an efficient model-based fitting algorithm. This software package is freely available under the GNU public license and can be found at the website [8]. A screen shot of the program is given in Fig. 1.

2. Model-based parameter estimation and the efficient model estimation algorithm

For atomic resolution (scanning) transmission electron microscopy ((S)TEM) images, statistical parameter estimation theory provides an excellent tool to quantitatively extract unknown structure parameters [9, 10]. In these atomically resolved images, the intensity is peaked at the

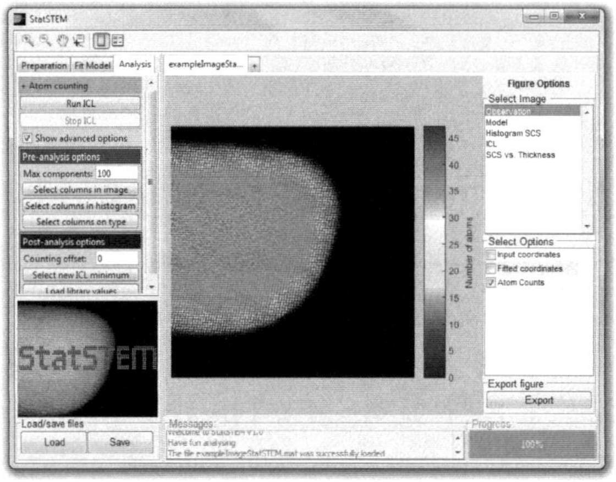

Figure 1. A screen shot of the StatSTEM program, showing atom-counting results on an experimental image of Au nanorod.

atomic column position and can be modelled as a Gaussian function [11, 12]. The expectation of the intensity of pixel (k,l) at position (x_k, y_l) can then be described by an expectation model $f_{kl}(\boldsymbol{\theta})$ with $\boldsymbol{\theta}$ the vector of unknown structure parameters, as:

$$f_{kl}(\boldsymbol{\theta}) = \zeta + \sum_{i=1}^{I} \sum_{m_i}^{M_i} \eta_{m_i} \exp\left(-\frac{\left(x_k - \beta_{x_{m_i}}\right)^2 + \left(y_l - \beta_{y_{m_i}}\right)^2}{2\rho_i^2} \right) \quad (1)$$

with ζ a constant background, ρ_i the column dependent width of the Gaussian peak, η_{m_i} the column intensity of the m_ith Gaussian peak, $\beta_{x_{m_i}}$ and $\beta_{y_{m_i}}$ the x- and y-coordinate of the m_ith atomic column respectively. The index i refers to atomic columns of the same atom type with I different types and the index m_i refers to the mth column of type i with M_i the number of columns of type i. The model includes a constant background which can account for the black level of the detector or for a constant contribution of a sample support. The indices in the summation of Eq. (1) can be simplified in the case of a mono-type crystalline nanostructure since then only one column type is present.
The parametric model of Eq. (1) is fitted to the experimental data using the uniformly weighted least squares criterion quantifying the similarity between the experimental images and the model. A direct implementation of the least squares estimator, where all parameters are simultaneously estimated, is computationally very intensive and can only be applied when dealing with a limited number of projected atomic columns. Therefore, we proposed a new, efficient algorithm [7]. This new algorithm divides the image into smaller segments containing individual columns without ignoring the overlap of image intensities between neighbouring columns. This overlap is estimated in an iterative procedure by subtracting in each iteration step the estimated contributions of neighbouring columns from the small segment of the single atomic column.

3. Benefits of a model-based approach

In order to illustrate the benefits of this new method in a statistical manner, the accuracy and precision are evaluated for the estimated atomic column position and the total scattered intensity of an atomic column, i.e. the so-called scattering cross-section which equals the volume, $V = 2\pi\eta_{m_i}\rho_i^2$, under the estimated Gaussian peak and which can be used to quantify the chemical composition and the number of atoms. This evaluation is performed as a function of thickness for the central atomic column of a simulated ADF STEM image of Pt along the [100] zone axis

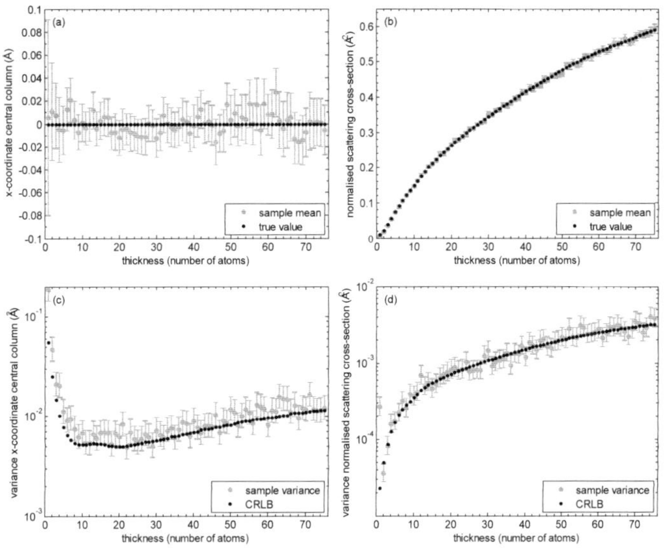

Figure 2. Accuracy and precision of the estimated position coordinate and scattering cross-section as a function of thickness in a Pt crystal for an incoming electron dose of $5 \cdot 10^2 e^- Å^{-2}$. (a,b) The true values and the sample mean together with the 95% confidence intervals. (c,d) The highest attainable precision, the CRLB, and the sample variance together with the 95% confidence intervals.

orientation. Using 100 noise realisations for an incident electron dose of $5 \cdot 10^2 e^- Å^{-2}$, the sample mean is compared with the true value in order to evaluate the accuracy. In order to evaluate the precision, the sample variance is compared with the theoretical lower bound on the variance, the so-called Cramér-Rao lower bound (CRLB). It is shown that even for these low electron dose images structure parameters are accurately measured with the highest attainable precision (Fig. 2).

Furthermore, the advantages of our model-based approach are illustrated as compared to the computationally less expensive approach where the overlap between neighbouring columns is neglected. This is first shown by estimating the spacing between two neighbouring Pt atomic columns of 5 and 3 atoms thick as a function of the distance (Fig. 3a) when taking overlap into account and when neglecting it (Fig. 3b). Secondly, a comparison between the estimated volumes and the integrated intensities from a Voronoi cell reveals that the estimated volumes give more accurate results for predicting the scattering cross-section (Fig. 3c). Therefore, this comparison further highlights the advantages of our model-based approach.

4. Conclusions

In conclusion, a new efficient algorithm is developed enabling a quantitative analysis of (S)TEM images having a large field of view. As this algorithm still accounts for the overlap of image intensities between neighbouring columns, it provides accurate measurements of structure parameters with the highest attainable precision. To provide end-users this powerful quantification method, a user-friendly software package is available.

Acknowledgments

The authors acknowledge financial support from the Research Foundation Flanders (FWO, Belgium) through project funding (G.0374.13N, G.0368.15N, G.0369.15N, WO.010.16N) and a

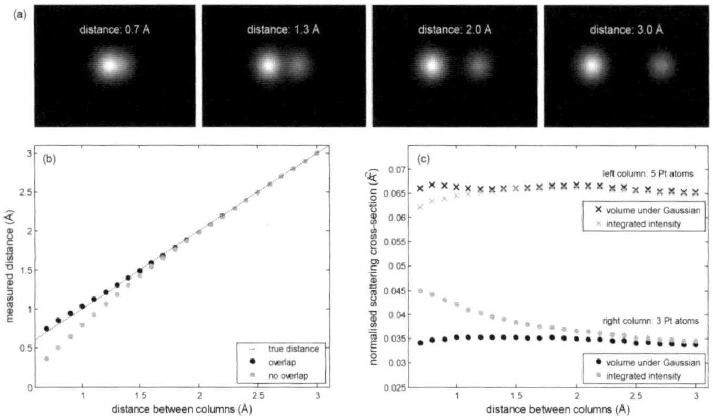

Figure 3. Influence of a neighbouring column on the estimated parameters. (a) Examples of simulated images of a 5 and 3 Pt thick column with a varying distance between the columns. (b) The estimated distance when taking the overlap of neighbouring columns into account and when neglecting it. (c) The estimated scattering cross-sections obtained from the volumes under the fitted Gaussian functions and from integrating intensities from Voronoi cells.

PhD research grant to K H W van den Bos, and a postdoctoral research grant to A De Backer. The research leading to these results has received funding from the European Union Seventh Framework Programme under Grant Agreement 312483 - ESTEEM2 (Integrated Infrastructure Initiative-I3). A Rosenauer is acknowledged for providing the STEMsim program.

References

[1] Van Aert S, Turner S, Delville R, Schryvers D, Van Tendeloo G and H S E K 2012 *Advanced Materials* **24** 523–527

[2] Van Aert S, Batenburg K J, Rossell M D, Erni R and Van Tendeloo G 2011 *Nature* **470** 374–377

[3] Van Aert S, De Backer A, Martinez G T, Goris B, Bals S and Van Tendeloo G 2013 *Physical Review B* **87** 064107

[4] De Backer A, Martinez G T, Rosenauer A and Van Aert S 2013 *Ultramicroscopy* **134** 23–33

[5] Van Aert S, Verbeeck J, Erni R, Bals S, Luysberg M, Van Dyck D and Van Tendeloo G 2009 *Ultramicroscopy* **109** 1236–1244

[6] Martinez G T, Rosenauer A, De Backer A, Verbeeck J and Van Aert S 2014 *Ultramicroscopy* **137** 12–19

[7] De Backer A, van den Bos K H W, Van den Broek W, Sijbers J and Van Aert S 2016 *Ultramicroscopy* **171** 104–116

[8] URL `https://github.com/quantitativeTEM/StatSTEM`

[9] den Dekker A J, Gonnissen J, De Backer A, Sijbers J and Van Aert S 2013 *Ultramicroscopy* **134** 34–43

[10] Van Aert S, De Backer A, Martinez G T, den Dekker A J, Van Dyck D, Bals S and Van Tendeloo G 2016 *IUCrJ* **3** 71–83

[11] Van Dyck D 2002 *Advances in Imaging and Electron Physics* **123** 105–171

[12] Nellist P D 2007 Scanning transmission electron microscopy *Science of Microscopy* vol 1 ed Hawkes P W and Spence J H C (Springer New York) chap 2, pp 65–132

Imaging with the Super-resolution Microsphere Amplifying Lens (SMAL) Nanoscope

Sorin Laurentiu Stanescu[1], Sébastien Vilain[1], Valerio Galieni[1], George Goh[2], Katarzyna Karpinska[3], Irina Barbolina[2], Alex Sheppard[1], Steve Wright[1], Wei Guo[4], Lin Li[4]

[1] LIG Nanowise Limited, Manchester Science Park, Manchester, M15 6SE, UK
[2] LIG Biowise Limited, Manchester Science Park, Manchester, M15 6SE, UK
[3] School of Biological Sciences, Faculty of Biology, Medicine and Health, The University of Manchester, UK
[4] Laser Processing Research Centre, School of Mechanical, Aerospace and Civil Engineering, The University of Manchester, UK
E-mail: sorin.stanescu@lig-nanowise.com

Abstract. The basic principle of the SMAL (Super-resolution Microsphere Amplifying Lens) imaging technology that we invented has its roots in the SMON (Submerged Microsphere Optical Nanoscope) pioneering work. SMAL is a novel objective lens whose front lens assembly contains a microsphere and is replaceable. We built a nanoscope prototype with nano XYZ scanning capability, which integrates a SMAL objective lens, allowing us to achieve super resolution imaging (70 nm – 90 nm). We have resolved large area scans (200 μm x 200 μm) by contactless sample scanning.

1. Introduction and aim

The resolution limit for conventional optical microscopy is approximately 200 nm within the visible light spectrum due to the far-field diffraction limit. It follows that conventional techniques are not suitable for imaging objects with structures smaller than this limit, for example live viruses (typically 5–150 nm, with some up to 300 nm) [5]. Other techniques have been used to image this kind of samples, like Transmission Electron Microscopy (TEM) and Scanning Electron Microscopy (SEM), that are often used to observe specially prepared dead virus structures at very high resolutions (10 nm) in vacuum. These techniques require complex sample preparation, and are not suitable for in vivo imaging and measurements (the electron beam damages living cells, viruses, etc.). Atomic force microscopes (AFMs) offer good imaging of small features by using a contacting probe. However, samples may be easily damaged by the AFM tip. Moreover, this technique does not offer a real image, but a reconstructed image, and the configuration plus acquisition time is tipically very long (up to hours). Stimulated Emission Depletion (STED) fluorescence optical microscopy is a recently established method for imaging cellular structures, bacteria and viruses beyond the diffraction limit, down to a resolution of 6 nm. STED offers a better resolution, but the sample also requires a complex preparation (fluorescent labelling), which may not always be suitable for imaging living organisms. The fluorescent imaging techniques give good results mainly for organic samples [4]. However, for high resolution, these techniques are confronted with the challenge of photo bleaching which limits the exposure time to tens of seconds. Microsphere super resolution imaging has been demonstrated using microspheres positioned between the objective lens and the sample using the SMON (Submerged

Microsphere Optical Nanoscope) technique [1]. The employed microspheres are typically in the order of 1-100 μm in diameter. The high spatial frequency information is carried by evanescent waves present at the boundary of two different media with different refractive indices. These waves decay exponentially within 1 μm, so they cannot be observed with conventional objective lenses. The use of microspheres allows to capture such waves and propagate them into the far-field. The basic principle of the SMAL (Super-resolution Microsphere Amplifying Lens) imaging technique that we invented has its roots in the SMON pioneering work. SMAL is a novel objective lens whose front lens assembly contains a microsphere and is replaceable. We built a prototype nanoscope with nano XYZ scanning capability which integrates a SMAL lens. The microsphere gives us the super resolution (down to 70 nm – 90 nm), while the piezo stage allows for autofocusing and arbitrarily large area scans (up to its travel range limit: 200 μm x 200 μm).

2. Experimental setup

We developed an original microscope comprising of: a regular optical microscope system from commercial available opto-mechanical components, a manual XYZ stage for coarse adjustment, a computer-controlled XYZ piezo stage, a SMAL objective lens, a metallic custom chassis and a protective enclosure. The schematic of the experimental setup is shown in Figure 3.

2.1. Objective lens

The setup consists of a regular 100x (NA = 1.24) oil immersion objective lens with a disk attachment on the top of its hemispherical front lens. We spincoated a uniform thin film of NOA®81 UV curable glue onto a glass disk, then we placed a $BaTiO_3$ microsphere (we tried different sizes in different setups) with a refractive index of 1.93 on the glue layer. Then we cured the attachment for 1 hour (Figure 1) with UV light. We calculated the total thickness of the attachment such that the microsphere is in the focal point of the objective lens. This thickness varies depending upon the focal distance of the objective lens, the diameter and refractive index of the microsphere, as well as the wavelength of the irradiating light. We confirmed the value of the thickness experimentally, by using a simple setup consisting of a microsphere positioned onto a sample and a micrometric Z adjustment to move the objective lens on top of it. By evaluating the sharpness of the image at different Z positions, it is easy to find the optimal distance between the lens and the sphere. Special care must be taken in aligning the microsphere with the optical axis before fixing it onto the glass disk. The position of the sphere can be evaluated in real time by observing the image acquired with the nanoscope. We designed SMAL with a replaceable front lens assembly to allow the replacement of the microsphere, in case it gets damaged by repeated contact with the sample.

2.2. The microscope

We mounted the SMAL objective lens (Figure 2) onto our custom made white light reflection microscope. The microscope includes two stages: a manual XYZ with micrometric stage to allow for coarse sample exploration and focusing, and a software-controlled XYZ piezo stage for cotrolling the sample scan and the autofocus. The best quality imaging was obtained with oil and water immersion.

2.3. The software

Typically, the field of view of SMAL is tiny when compared to a standard objective lens, and it is dependent upon the size of the microsphere. Moreover, the image can be affected by a slight pincushion distortion. In order to obtain an undistorted, large area image of the sample in super resolution, a scan process is necessary. This is achieved by using a piezo stage with nanometric accuracy and a custom software. The sofware controls the camera and the stage and automatically acquires several images in different positions of the sample, in a grid-like pattern. Each image is corrected for distortion and cut to preserve only the part with meaningful super resolution information (at the centre of the microsphere). Then, the software stitches all the images together to obtain a large area mosaic. Sometimes, a grid-like pattern may appear in the final mosaic, depending on a number of factors, including the material of the sample and the immersion medium. So, as an optional final step is the removal of such pattern by using standard image processing algorithms. The scan is typically fast when compared to super resolution imaging techniques like AFM. When using SMAL, a typical 20 μm x 20 μm scan of a silicon microprocessor sample with a 15 μm diameter microsphere requires stitching 784 images and takes 2 minutes and 16 seconds.

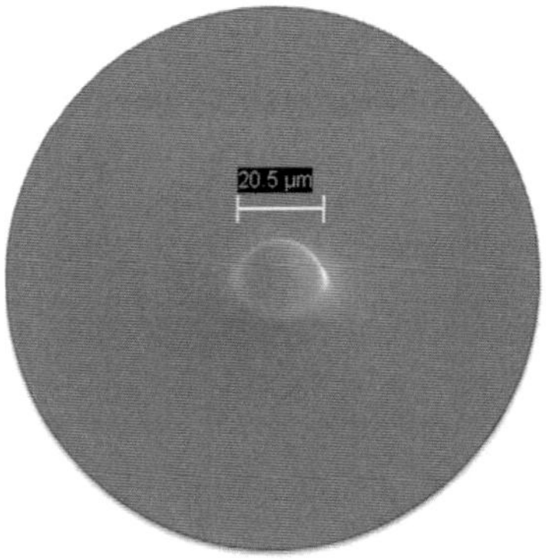

Figure 1. SEM imaging of our embedded microsphere in spin coated UV curable glue (20 μm diameter barium titanate microsphere)

Figure 2. Front side rendering (not to scale) of the proposed objective lens with the replaceable front lens assembly

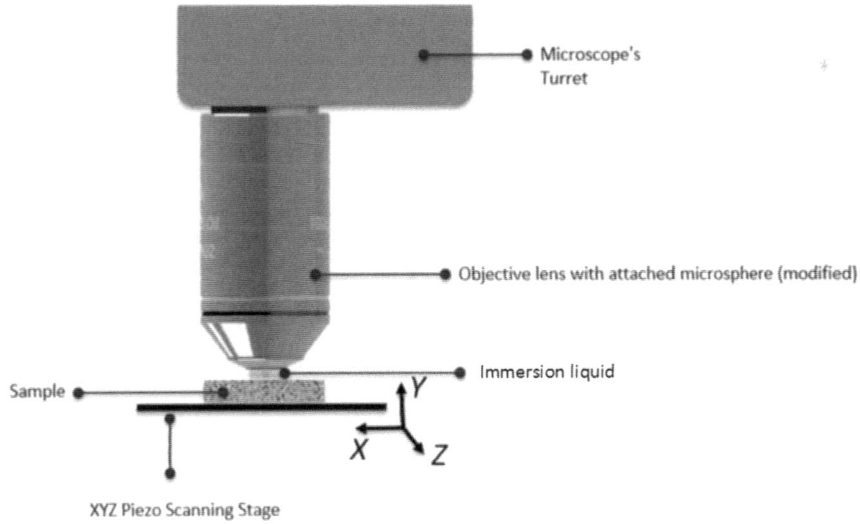

Figure 3. Experimental Setup using SMAL

Our software suite allows the user to take large area scans up to 200 μm x 200 μm (the travel range of the stage) with a lateral resolution down to 70 nm, depending on the sample. Figure 4 (a) shows the scanning of a 80 μm x 80 μm of an Intel® Core i3-530 "Clarkdale" microprocessor. Figure 4 (b) highlights a zoomed in area of the same sample where it is possible to observe features with sizes between 70 nm and 90 nm, as well as different circuit copper layers in depth.

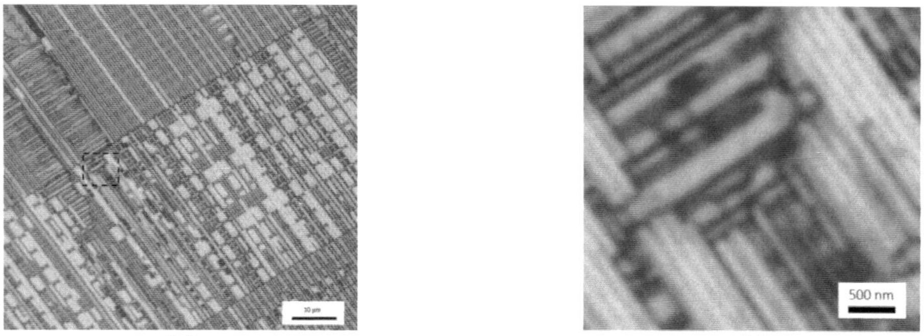

(a) (b)

Figure 4 – (a) 80 μm x120 μm scan of an Intel® Core i3-530 "Clarkdale" microprocessor using our SMAL based nanoscope; (b) zoomed in area of the same image

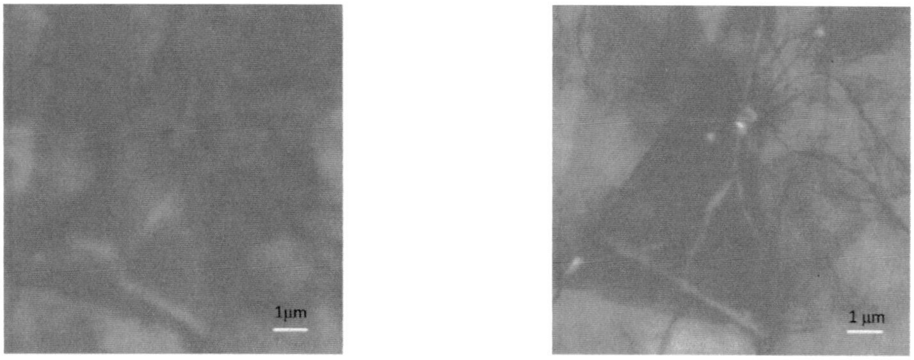

(a) (b)

Figure 5 – (a) 100x, 1.24NA oil immersion objective lens imaging of a graphene sample; (b) the same sample scanned with our SMAL based nanoscope

3. Conclusion:

- With SMAL, we resolved features down to 70 nm - 90 nm of an Intel® Core i3-530 microprocessor;
- We could automatically scan arbitrarily large areas in super resolution;
- The technique is non-contacting, non-invasive, and it works naturally in both wet and dry media;
- This work opens new opportunities for the study of nanostructures, and is suitable for efficient quality control, as well as studying the interaction between cells/viruses/bacteria/drugs;
- In comparison with expensive equipment like SEMs and AFMs, which can only image the sample's surfaces, our method can see inner layers of materials with resolution beyond diffraction limit at a depth down to 3 μm (Figure 4);
- Unlike other imaging technologies previously mentioned, colour imaging with super resolution can give an estimate of the thickness of graphene layers (Figure 5b).

References

[1] Li L, Guo W, Yan Y, Lee S and Wang T 2013 *Light Sci. Appl.* 2(9), e104
[2] Lee S and Li L 2015 *Opt. Commun.* 334, 253–257
[3] Wang Z, Guo W, Li L, Luk'yanchuk B, Khan A, Liu Z and Chen Z 2011 *Nature Comm.* 2, 218

Optical ptychography with extended depth of field

Peng Li and Andrew Maiden

Department of Electronic and Electrical Engineering, University of Sheffield, Sheffield, UK, S1 3JD.

peng.li.shef@sheffield.ac.uk

Abstract. Ptychography is an increasingly popular phase imaging technique. However, like any imaging technique it has a depth of field that limits the volume of a thick specimen that can be imaged in focus. Here, we have proposed to extend the depth of field using a multislice calculation model; an optical experiment successfully demonstrates our proposal.

1. Introduction

Ptychography is a coherent diffraction imaging technique that can deliver quantitative phase information at diffraction limited resolution [1]. Its conventional implementation requires a coherent illumination of finite extent to be scanned over a specimen at a raster grid of positions and the corresponding diffraction patterns to be recorded by a camera somewhere downstream [2]. The illuminated area between adjacent positions needs to partially overlap, such that redundant information exists in the recorded diffraction patterns. Since only the intensity is detected (the phase information is lost), a phase retrieval algorithm is needed to re-phase the recorded diffraction patterns and reconstruct the complex transmission function of the specimen.

However, ptychography has a limited depth of field; it requires the specimen to be thin enough such that the wavefront that exits the specimen (the exit wave) is accurately approximated by the product of the illumination function and the transmission function of the specimen [2, 3], as shown in Fig. 1a. Under the Born approximation, the depth of field of ptychography is given by Eq. (1) and explained in Fig. 1b [4]:

$$\mathrm{DOF}=\frac{1}{k_z}=\frac{\lambda}{2-\sqrt{1-\mathrm{NA}_I^{\,2}}-\sqrt{1-\mathrm{NA}_S^{\,2}}}. \qquad (1)$$

Here k_z is the accessible volume along optical axis in reciprocal space, λ is the wavelength of the illumination, NA_I is the numeric aperture of the illumination and NA_S is the numeric aperture of the specimen.

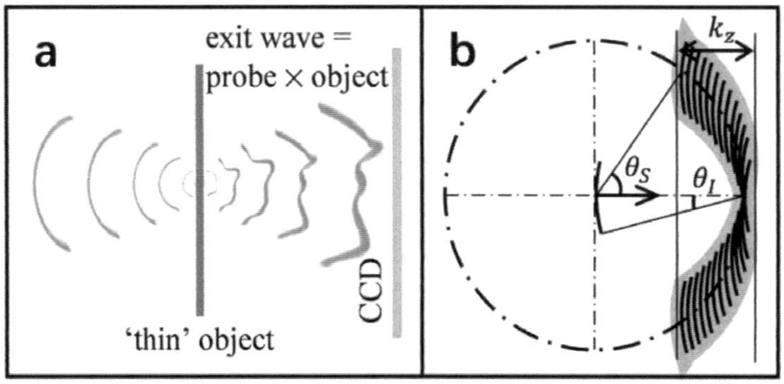

Figure 1. The exit wave formation model for a thin object. (a) The schematic diagram of the exit wave formation of a thin object. (b) The Ewald sphere construction of the configuration in (a). θ_S is the half angle of the specimen scattering angle and θ_I is the half angle of the illumination angle.

When the specimen thickness exceeds this limit, the approximation breaks down and conventional phase retrieval algorithms do not give accurate reconstructions. A computational sectioning model called the multislice method, widely used in electron microscopy [5], can be adopted to circumvent this problem. The method computationally sections a thick specimen into a set of thin slices, each of which allows the formation of its exit wave using the multiplicative approximation [3], as shown in Fig. 2. With modifications of the conventional phase retrieval algorithms [3], all the slices can be reconstructed, giving a three dimensional representation of the specimen.

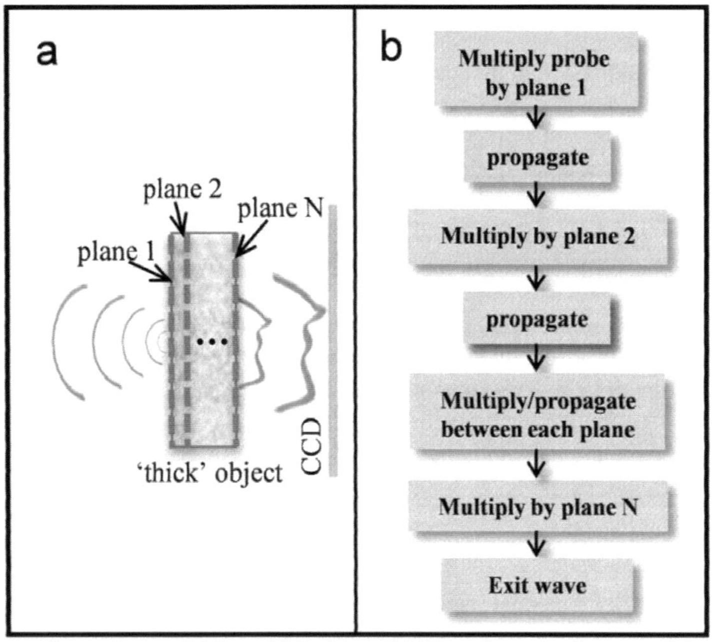

Figure 2. The exit wave formation model for a thick object. (a) The schematic diagram of the exit wave formation of a thick object. (b) The computational model of the multislice calculation.

Instead of pursuing the 3D imaging performance of multislice ptychography, here we demonstrate its ability to extend the depth of field via an optical experiment. This has been demonstrated in x-ray regime [6], although in this case the breakdown of the multiplicative approximation is nowhere near as severe as that in the visible light regime, due to the strong penetration power of x-rays.

IOP Conf. Series: Journal of Physics: Conf. Series **902** (2017) 012015 doi:10.1088/1742-6596/902/1/012015

2. Experiment and results

The optical setup is illustrated in Fig. 3a. The illumination source is a 635nm laser diode. The ptychographic scans consisted of a 15×15 raster grid of specimen positions with a nominal 20μm step size, plus ±20% random offsets. The confined illumination was formed by focusing a collimated laser beam via a lens of 35mm focal length. A piece of plastic film was used to increase the angular range of the illumination. An objective lens with a NA of 0.65 combined with a tube lens of 100mm focal length magnified the near-field diffraction pattern onto a CCD detector. The calibrated magnification for our system was 21. The CCD detector had 2048×2048 pixels with a pitch of 7.4μm^2. The distance from the specimen to the virtual CCD plane was set at 0.5mm. The recorded diffraction patterns were binned by a factor of 2 before processing, and an example is shown in Fig. 3b.

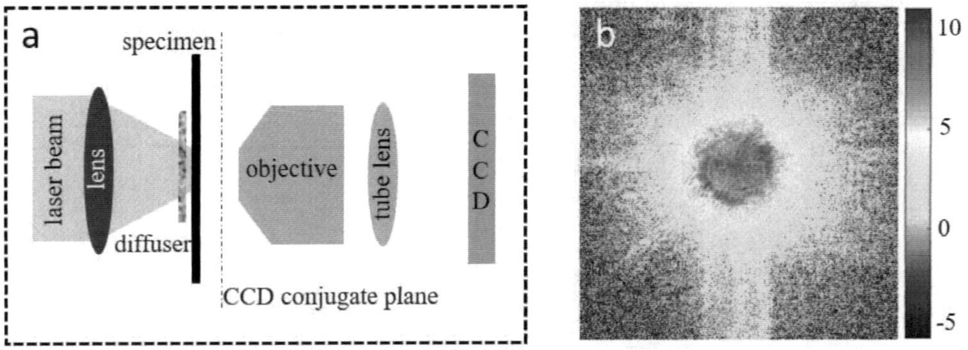

Figure 3. The optical experiment. (a) The experimental configuration. (b) An example diffraction pattern.

The reconstruction was performed by scaling all the diffraction patterns down onto the virtual CCD plane. The propagation between the specimen and the virtual CCD plane was modelled using the angular spectrum propagator [7]. The pixel size of the specimen was equal to the pixel size of the virtual CCD, namely 2×7.4/21=0.70μm. For comparison, an incoherent light microscope image of the specimen (a microscope slide of honeybee mouth parts), taken with a 10 × objective (0.25 NA) and focused at the bottom surface of the specimen, is given in Fig. 4a. As we can see, the specimen is quite thick, so there are many out of focus features – this results in a 2D image which is difficult to interpret, because the out of focus light masks the in focus structures.

For the ptychographic imaging, first we tried a reconstruction assuming a single slice in the specimen, which gave the phase reconstruction shown in Fig. 4b. The reconstruction did not completely fail due to the weak scattering of the specimen, however it is blurred and the fine features, such as the bee's hairs, are invisible. Next, we used four slices with equal separations of 33μm in a second reconstruction using the multislice algorithm described in [3], which resulted in the images shown in Fig. 4c. Features at different axial positions are now separated in different slices, although there is still cross-talk between slices. To obtain the 2D image with extended depth of field, we summed up the four phase images, as shown in Fig. 4d. Compared to the single slice reconstruction, the extended depth of field image enables all the features in a thick specimen to be nicely in focus.

3. Conclusions

In this work, we have demonstrated the potential of multislice ptychography in extending the depth of field via an optical experiment. It enables ptychography to image a thick specimen with all the features in focus.

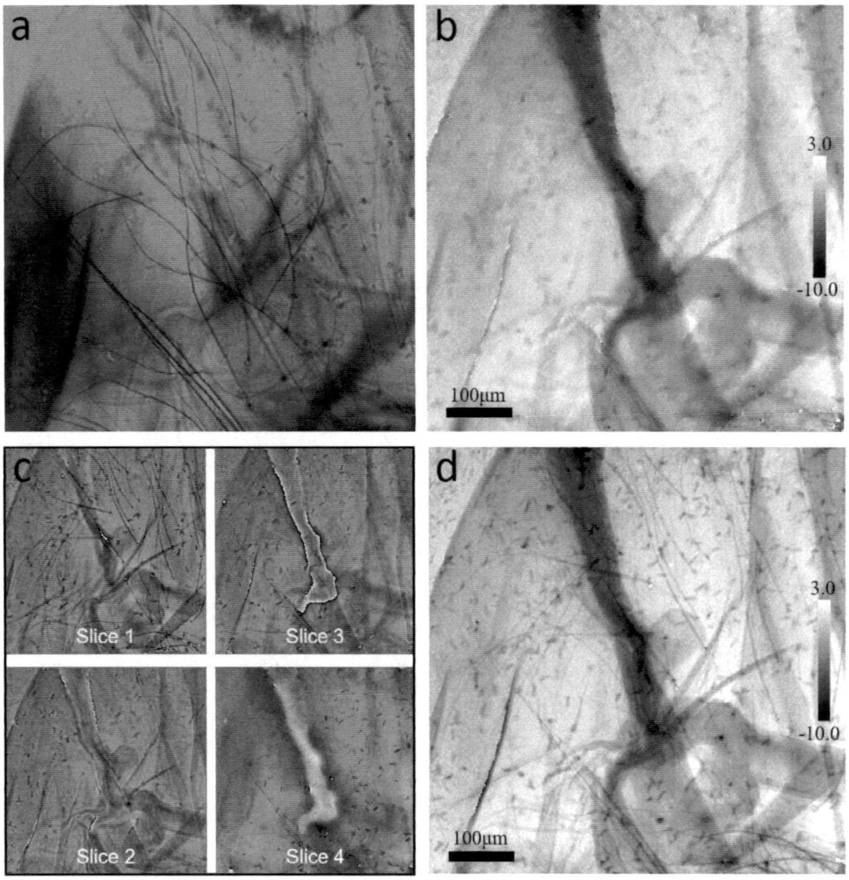

Figure 4. The specimen reconstruction results. (a) The incoherent light microscope image focused on the bottom surface. (b) The ptychographic reconstruction assuming single slice of the specimen. (c) The reconstruction of multislice ptychography using four slices with a constant separation of 33μm. (d) The extended depth of field image by summing up the four slice reconstructions.

References

[1] Rodenburg J M 2008 *Advances in imaging and electron physics* (Elsevier)
[2] Thibault P, Dierolf M, Menzel A, Bunk O, David C and Pfeiffer F 2008 *Science* **321** 379-82
[3] Maiden A M, Humphry M J and Rodenburg J M 2012 *J. Opt. Soc. Am.* A **29** 1606-14
[4] Tian L and Waller L 2015 *Optica* **2** 104-11
[5] Cowley J M and Moodie A F 1957 *Acta Crystallogr.* A **10**, 609-19
[6] Tsai E H, Usov I, Diaz A, Menzel A and Guizar-Sicairos M 2016 *Optics Express,* **24** 29089-108
[7] Goodman J W 2005 *Introduction to Fourier optics* (Roberts and Company Publishers)

IOP Conf. Series: Journal of Physics: Conf. Series **902** (2017) 012016 doi:10.1088/1742-6596/902/1/012016

Quantification of Fe-oxidation state in mixed valence minerals: a geochemical application of EELS revisited

A P Brown[1], S Hillier[2] and R M D Brydson[1]

[1] School of Chemical and Process Engineering, University of Leeds, Leeds, LS2 9JT, UK
[2] The James Hutton Institute, Craigebuckler, Aberdeen, AB15 8QH, UK

Abstract. The use of TEM and electron energy loss spectroscopy to determine levels of Fe^{3+} in solids at nanometre spatial resolution is well established. Here we assess this technique at the energy resolution of a monochromated electron source (0.3 eV) and with the capability of simultaneously acquiring core- and low-loss spectra for absolute energy calibration of core-loss chemical shifts. Fe $L_{2,3}$-edges of four oxide standards give a characteristic energy loss near edge structure (ELNES) and the L_3-peak maxima of the spectra exhibit an energy shift of ~ 1.8 eV between ferrous (709 eV) and ferric (710.8 eV) iron. We show that diagenetic chlorites present in iron oxide loaded 'red beds' contain predominately Fe^{3+}, based on the absolute energy position of the low signal Fe L_3-edges (centred around 710.8 eV). We also fit our standard-spectra to Fe $L_{2,3}$-edges from magnetite and maghemite nanoparticles to investigate iron site occupancy. An additional chemical shift needs to be applied to the octahedral ferrous iron site component to achieve a plausible, linear fit to the magnetite spectrum which we speculate may be due to electron hopping between the mixed valence octahedral iron sites in magnetite.

1. Introduction

Geochemical applications of Electron Energy-Loss Spectroscopy (EELS) in the Transmission Electron Microscope (TEM) have shown a wide range of benefits, including elemental quantification, the determination of oxidation state, and local co-ordination of atomic species [1]. Of these, the determination and quantification of Fe oxidation state is of particular geochemical significance in terms of the information it provides on redox, fugacity, pressure and temperature conditions at crystallisation or alteration.

The TEM-EELS methods of Garvie and Buseck [2] and van Aken, Liebscher and Styrsa [3] allow quantification of ferric iron to total iron in a range of silicates and oxides with a total iron content of a few atom % or more. This work revisits similar analyses, originally reported by Pan et al. [4], but with the improved energy resolution of a monochromated electron source and additional, absolute edge energy measurement by dual EELS acquisition [5].

2. Methods

We present data from the iron valence and co-ordination standards used by Pan et al. [4] to provide a reference and then investigate measuring valence in low iron content diagenetic chlorites (provided by Stephen Hillier from Old Red Sandstone of the UK [FWW-5 and ED2], the Permian of Poland [ROT-1] and the Triassic of the North Sea [CHL-25 and -34]; all sieved to < 2 μm clay size fractions) and iron site occupancy in synthetic iron oxide nanoparticles (made in-house or purchased from Alfa-Aesar). We have used a FEI Titan[3] Themis 300 kV G2 S/TEM at the University of Leeds which is equipped with a monochromator (tuned to 0.3 eV energy resolution), FEI SuperX EDX detectors, a Gatan Quantum ER 965 imaging filter (with Dual EELS; this enables energy calibration of a core-loss edge with a simultaneously acquired low loss spectrum) and a Gatan OneView CCD. Image coupled EEL spectra (~ 6.6 mrad collection and ~ 1.0 mrad convergence semi-angles) were acquired in TEM diffraction (using the smallest selected area electron diffraction aperture of ~ 200 nm diameter at the image plane) from crushed dispersions of minerals drop-cast on holey carbon films (Agar Scientific Ltd). Spectra were energy calibrated (to the simultaneously acquired low-loss) and background subtracted (power law) in Gatan's GMS software (version 3.1), t/λ values were generally < 0.35.

3. Results

3.1. Iron Oxide Standards

Fe $L_{2,3}$-edges of the four oxide standards exhibiting different iron valence and co-ordination give characteristic energy loss near edge structure (ELNES). Mineral details and EEL spectra recorded at 0.8 eV energy resolution are reported in Pan et al. [4]. Here, there are clear shoulders at high and low energies of the main L_3 peaks for ferrous and ferric iron respectively and the peak maxima of the spectra exhibit a chemical shift of ~ 1.8 eV between ferrous (709 eV) and ferric (710.8 eV) iron valence states (Figure 1). The energy-loss scale is now absolute due to the Dual EELS capability of the spectrometer. There was little change to the mineral morphologies or Fe edge shapes over the 30 s acquisition times (up to 1 min total exposure; at a screen current of 20 nA prior to insertion of the SAED aperture).

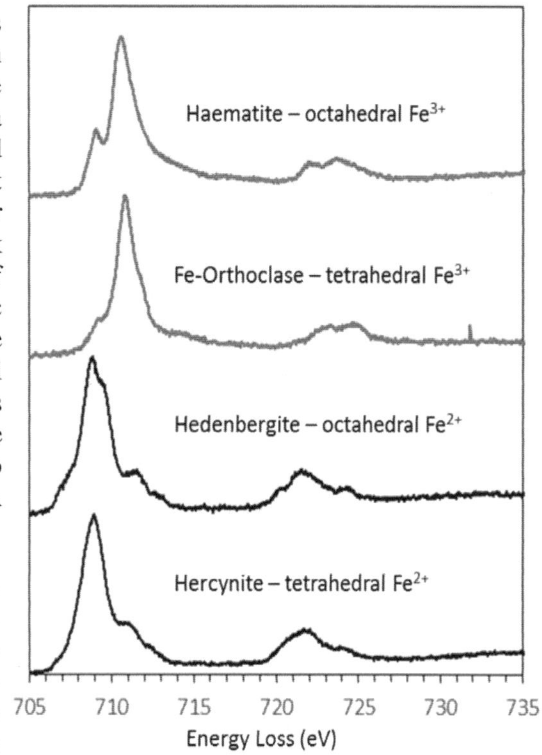

Figure 1. Reference Fe $L_{2,3}$-edge energy loss spectra; from bottom up, hercynite (tetrahedral coordinated ferrous iron), hedenbergite (octahedral coordinated ferrous iron), Fe-orthoclase (tetrahedral coordinated ferric iron) and haematite (octahedral coordinated ferric iron).

3.2. Dioctahedral Chlorites from Red Beds

Diagenetic chlorites are a common component of many sedimentary rocks. In most settings aluminous dioctahedral chlorites are rarer than ferro-magnesian trioctahedral types, however in some iron oxide-rich 'red beds' this trend may be reversed. It is hypothesised here that early formed hematite may play a crucial role as a buffer on the oxidation state of Fe in some red bed diagenetic systems. Under such conditions, it is postulated that the formation of dioctahedral chlorites (sudoite and tosudite) is favoured over that of the more common Fe^{2+} rich trioctahedral relatives because they can readily and preferentially accommodate Fe^{3+} [6].

Three dioctahedral chlorite samples provided by Hillier have clay particles present that are some type of sudoite (FWW5 and ROT-1) or tosudite (ED2) by powder X-ray diffraction (data not shown). Chlorites consist mainly of Si, Al, Fe and Mg and the presence of significant amounts of K, Na and Ca would indicate contamination with other clay minerals. Two reference chlorites, a tri- and dioctahedral form (CHL-25 and CHL-34) are also investigated and these are expected to be ferrous and ferric bearing respectively. Fe $L_{2,3}$-edges taken from the sudoite (FWW5 and ROT-1) and tosudite (ED2) samples have low signal but are centred around 710.8 eV energy loss, indicating that they contain predominately Fe^{3+} while the tri- and dioctahedral chlorite references (CHL-25 and CHL-34) are mixed valence and Fe^{3+}-rich respectively (Figure 2).

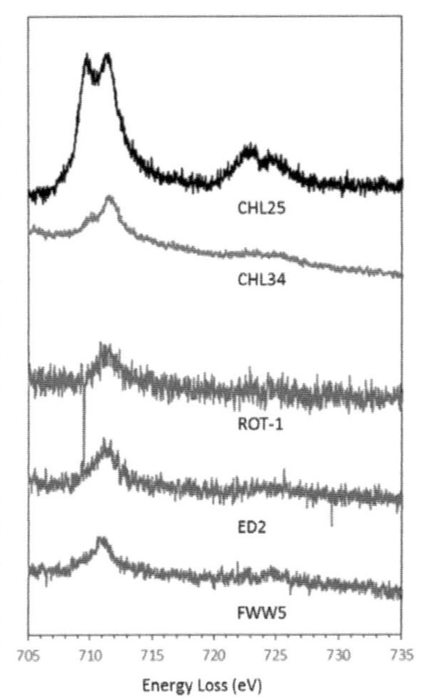

Figure 2. Fe $L_{2,3}$-edges of particles from the chlorite standards and unknowns. Iron levels are sufficiently low to make valence determination from Fe-ELNES difficult however the absolute edge energies of the peak maxima permit an estimate. The tri-octahedral reference chlorite (CHL-25) is iron-rich and mixed valence whereas the dioctahedral reference (CHL-34) is lower in iron content and Fe^{3+}-rich (the Fe L_3 peak maxima is at 710.8 eV). The sudoites (FWW5 and ROT-1) and tosudite (ED2) have even lower iron content but are clearly all Fe^{3+}-rich (given that they too have Fe L_3 peak maxima centred around 710.8 eV). EDX spectroscopy of the unknowns was used to confirm the chlorite-like composition of the particles tested by EELS.

3.3. Fe_3O_4 (magnetite) and γ-Fe_2O_3 (maghemite)

Nanoparticulate forms of Fe_3O_4 (magnetite) and γ-Fe_2O_3 (maghemite) were synthesized in-house (following the method described in Brown et al. [7]) or purchased from Alfa-Aesar respectively. Fe_3O_4 and γ-Fe_2O_3 have the same crystal structure however the former contains mixed valence iron and the latter has only Fe^{3+} (with corresponding octahedral iron site vacancies to achieve stoichiometry). The unit-cell lengths of pure Fe_3O_4 and pure γ-Fe_2O_3 are given as 0.839 and 0.835 nm respectively (for example, Zhu et al., [8]).

Powder X-ray diffraction of the materials assessed here confirms both samples have the expected inverse spinel structure and Rietveld refinement suggests the unit-cell length of the magnetite sample is 0.838 nm and the maghemite sample is 0.833 nm (data not shown). The Fe $L_{2,3}$-ELNES of these samples (Figure 3, black spectra) are consistent with Fe $L_{2,3}$-X-ray Absorption Spectra reported by Zhu et al. [8].

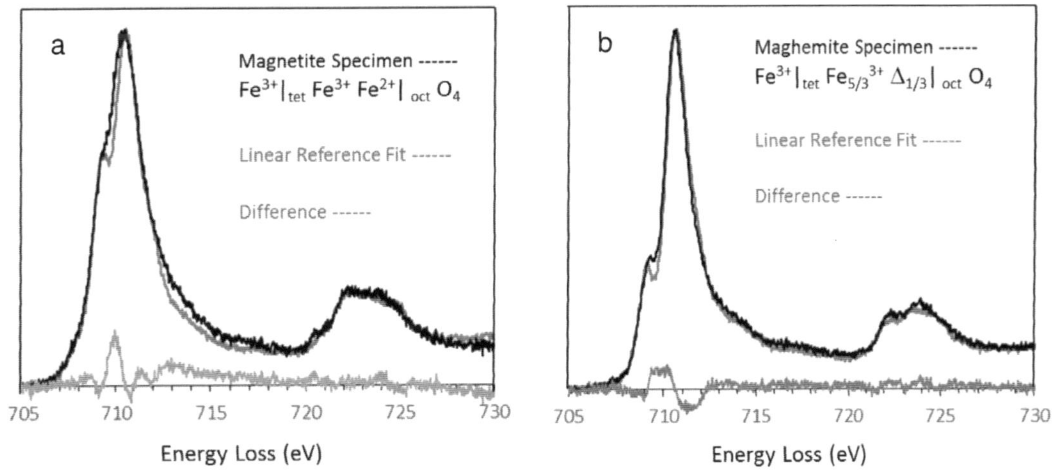

Figure 3. Fe $L_{2,3}$-edges of magnetite (a) and maghemite (b) samples (black). Also shown are linear reference fits (red) and difference spectra (grey) obtained by combining the previously measured iron valence and site standards (Figure 1) in the intensity ratios indicated by the formulae given in the figure legend and subtracting these from the experimental spectra.

IOP Conf. Series: Journal of Physics: Conf. Series **902** (2017) 012016 doi:10.1088/1742-6596/902/1/012016

Linear reference fits to these energy loss spectra have been achieved by adding the appropriate, intensity normalised standard spectra (Figure 1) in the intensity ratios indicated by the formulae for stoichiometric magnetite and maghemite (given in figure legends, Figure 3). The magnetite fit spectrum (1:1:1 for tetrahedral ferric: octahedral ferric: octahedral ferrous iron), also requires the octahedral Fe^{2+} component to be given a 0.8 eV offset to higher energy loss whereas the maghemite fit is achieved with no adjustment in energy loss of the two components (1:5/3 for tetrahedral ferric : octahedral ferric iron); The required offset for the octahedral Fe^{2+} component in magnetite may be a result of the strong asymmetry of the hedenbergite standard spectrum (i.e., changes in octahedral site symmetry for ferrous iron in hedenbergite and magnetite) or could be due to electron hopping that is known to occur between the octahedral coordinated iron sites in magnetite (e.g., [9]).

4. Conclusion

Using TEM and electron energy loss spectroscopy to determine Fe^{3+} content of minerals has been revisited at the energy resolution of a monochromated electron source (0.3 eV) and with the ability to simultaneously acquire core- and low-loss spectra for absolute energy calibration of core-loss chemical shifts. We show a reproducible ~1.8 eV chemical shift between ferrous (709.0 eV) and ferric (710.8 eV) Fe L_3 core-EELS edges. This absolute edge energy position is then used here to identify the dominance of Fe^{3+} in low iron content diagenetic chlorite minerals. In addition, Fe $L_{2,3}$-edges from magnetite and maghemite nanoparticles have been acquired and fitted by our iron-site standard spectra to investigate iron site occupancy. We find an additional chemical shift needs to be applied to the octahedral ferrous iron site component to achieve a plausible, linear fit to the magnetite spectrum and we speculate that this may be due to electron hopping between the mixed valence octahedral iron sites in magnetite however this finding requires further investigation.

References

[1] Brydson RMD, Brown AP, Benning LG and Livi KJT 2014 *Revs in Min and Geochem.* Analytical Transmission Electron Microscopy 78 219-269.

[2] Garvie LAJ and Buseck PR 1998 *Nature* Ratios of ferrous and ferric iron from nanometer-sized areas in minerals 396 667-670.

[3] van Aken PA, Liebscher B and Styrsa VJ 1998 *Physics and Chemistry of Minerals* Quantitative determination of iron oxidation states in minerals using Fe L-2, L-3-edge electron energy-loss near-edge structure spectroscopy 25 323-327.

[4] Pan YH, Vaughan G, Brydson R, Bleloch A, Gass M, Sader K and Brown A 2010 *Ultramicroscopy* Electron-beam-induced reduction of Fe^{3+} in iron phosphate dihydrate, ferrihydrite, haemosiderin and ferritin as revealed by electron energy-loss spectroscopy 110(8) 1020-1032.

[5] Gubbins A, Barfels M, Trevor C et al. 2010 *Ultramicroscopy* The GIF Quantum, a next generation pots-column imaging filter 110(8) 962-970.

[6] Hillier S 2003 *Encyclopaedia of sediments and sedimentary rocks* Chlorite in Sediments Kluwer 123-127

[7] Brown AP, Moore RGC, Evans SD and Brydson R 2001 *J. Phys. Conf. Ser.* Characterisation of iron oxides nanoparticles using EELS 168 255-258.

[8] Zhu X, Kalirai SS, Hitchcock AP and Bazylinski DA 2015 *J. of Elec. Spec. and Rel. Phenom.* What is the correct Fe L_{23} X-ray absorption spectrum of magnetite? 199 19-26.

[9] Rosencwaig C 1969 *Can. J. Phys.* Double exchange and electron hopping in magnetite. 47 2309-2317.

Acknowledgements

We gratefully acknowledge grant funding for the TEM from EPSRC (EP/M028143/1) and additional thanks to Dr Jenny Forrester, SCaPE, Leeds for the XRD analyses of the magnetite and maghemite specimens.

Characterization of a Miniature Electron Energy Analyzer for Scanning Electron Microscopes.

Ashish Suri[1], Andrew Pratt[2], Steve Tear[2], Christopher Walker[2] and Mohamed El-Gomati[1]

[1] Department of Electronics, University of York, York, UK
[2] Department of Physics, University of York, York, UK
ashish.suri@york.ac.uk

Abstract. We report the design and experimental characterisation of a miniature detector for the scanning electron microscope based on the Bessel Box (BB) electron energy analyser which has a simple cylindrical geometry. We report on the simulation and operation of a prototype BB. The energy resolution of a single BB has been numerically calculated and experimentally characterised to be < 1%. This miniature electron detector is designed to be used close to the sample, alleviating the effects of the ambient electrostatic and magnetic fields.

1. Introduction
Advancements in scanning electron microscopy (SEM) have been one of the major nanotechnology enablers. A magnified image of the sample is generated in the SEM by bombarding it with an electron beam and detecting the electrons that scatter off the surface along with the electrons that are generated in the sample [1]. In conventional SEMs, the generated secondary electrons (SEs) are detected by the widely adopted Everhart-Thornley detector via a positively biased input-grid [2]. However, in doing so, the energy and angular information of the SEs are lost. This detector also collects other tertiary electrons that are generated from the specimen chamber walls by energetic backscattered electrons, thus reducing the signal quality obtained. Preserving the energy and angular information of the SEs is crucial to interpret the SEM image of the sample under study. Thus, a small-scale detector is required that can preserve this information and the performance of which is little affected under the influence of ambient electrostatic and magnetic fields inherent to electron microscopes. In this work we propose the use of the lesser known Bessel box (BB) electron energy analyser (EEA) [3]. We examine the operation of the Bessel box using SIMION 8.1 [4] simulations and experimentally test a prototype.

The BB has a very simple cylindrical geometry consisting of 3 electrodes: a central cylinder and two end caps/plates with input and output apertures (figure 1). The electric field generated by the voltages on these electrodes allow the BB to act as a band pass filter. Allen et al. [5] used a BB with annular ring input and output apertures whilst grounding the input and output electrodes. The major drawback of this design was the existence of multiple electron trajectories within the analyser. This problem was addressed by Stalder et al. by introducing a stopper on the optical axis inside the BB [6]. An alternative solution was recently introduced by Schwietz et al. by negatively biasing the output electrode at the same potential as the central cylinder electrode. This suppressed the higher modes from being detected and only the trajectories that are cylindrically symmetric can be detected [3].

2. Simulations

2.1. The Bessel Box
In this work we operate the BB in the retarding configuration very similar to Schiwietz et al. [3], wherein the central and output electrodes are shorted and negatively biased. The input electrode is grounded. The band pass filtering action is demonstrated in figure 1. A primary beam with a conical

distribution (half angle = 14°), is incident at the input electrode, whereby electrons experience the retarding radial and axial fields inside the BB. Electrons with energies less than the BB energy are repelled back as they are unable to overcome the energy barrier of the BB axial fields (figure 1(b)). This field decreases from a maximum at the central BB electrode walls to a minimum at the BB optic axis, thus causing the focussing action. For electron energies greater than the peak pass energy, the BB is unable to focus these at the output aperture, and only the on-axis electrons are detected as shown in figure 1(c). Electrons with energies closer to the peak pass energies are focussed at the output aperture (figure 1(a)) and can be detected at the collector plate. Simulations are carried out for a BB with internal diameter of 3 mm, length 6 mm and i/o aperture diameters which are 10 % of the BB diameter.

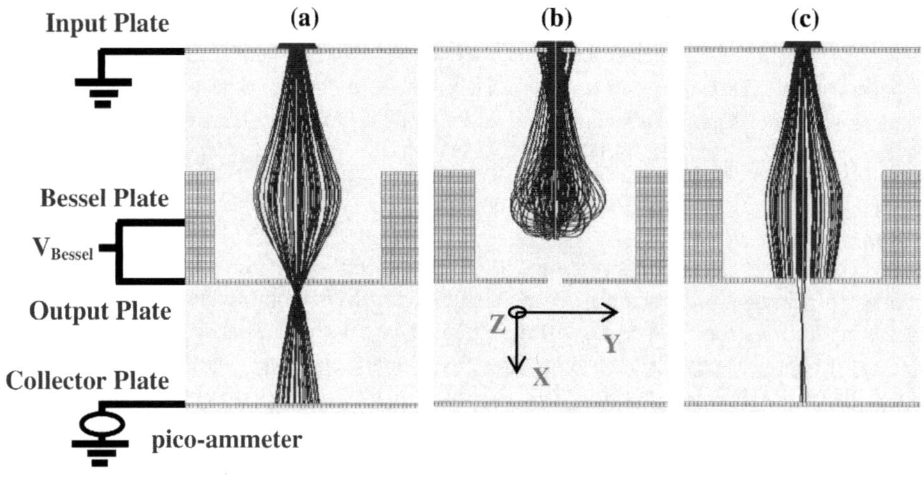

Figure 1. Simulation of electron trajectories in the BB. (a) For a given set of voltages the BB focusses electrons at the output aperture which are then collected. (b) Lower energy electrons have insufficient energy to reach the collector. (c) The BB is not able to focus higher energy electrons, however on-axis electrons do pass through.

2.2. Physics of operation: dispersion

The energy dispersion in the analyser leads to electrons being either stopped by the BB walls or collected through the output electrode (figure 2(a)). The electric field retards the electron velocity until it reduces to zero. To collect electrons within a narrow energy band, one has to define an output aperture. In this study the output aperture is placed on the optic axis at the output electrode. Figure 2(a) shows simulations of electrons entering at 11.2° to the optic axis and their energy is varied from 10.01 eV to 12 eV in equal steps. The plot of Y-coordinates on the output electrode with respect to the energy (figure 2(b)) enables calculation of the energy resolution ($\Delta E/E*100$). For the output aperture which is 10% of the BB diameter, the pass-band energy window is calculated to be 0.22 eV, which gives a resolution of 2%. This can be improved to 0.95% by reducing the output aperture to 5% of the BB diameter. However, this results in the trade-off between the energy resolution of the analyser and its collection efficiency.

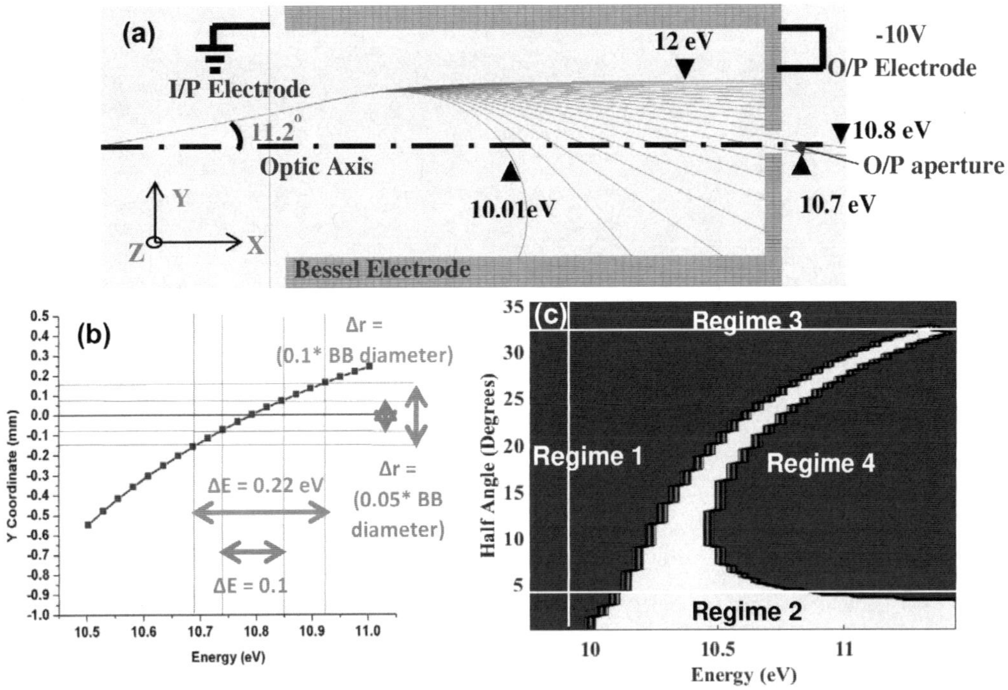

Figure 2. (a) The BB disperses the energies of the incident electrons. (b) The definition of the output aperture diameter results in the trade-off between collection and the resolution of the device. (c) The dispersion depends on the incident angle: yellow = electron detected, blue = no detection.

2.3. Physics of operation: angular dependence

So far the BB performance is assessed when the primary beam does not have any angular dispersion, which means all electrons are incident at the same angle. Practically this is never the case. An energy-angle convoluted surface-plot in figure 2(c) shows the detected electrons for a given a set of energies and angles, and the plot is divided into 4 regimes. Regime 1: Corresponds to lower energy electrons than the BB energy. Electrons with energies and angles within this regime are repelled back because axial velocity reversal takes place. Regime 2: The BB is unable to alter the on-axis electron trajectories as a result of which they go undeflected and are collected by the detector. For the BB to act as band-pass filter, the on-axis must be avoided. Regime 3: Corresponds to electron angles which are blocked by the geometrical design. Regime 4: This regime shows the focusing regime of the BB. There are two important points to be noted: firstly, electrons detected at higher incident angles are of higher energy, and secondly, the band-pass energy decreases with the increase in the incident angle.

3. Experimental Characterisation

A prototype BB was fabricated using 10mm x 10mm aluminium metal plates, which have thickness 5mm for the central cylinder and 0.2mm for the end caps and collector. The plates were biased using a high voltage operational amplifier (Kepco OPS 2000) controlled by a low voltage DC power supply (Keithley 2460). The BB was placed into a dedicated ultra-high vacuum chamber furnished with a hot cathode electron gun to directly illuminate the BB with a range of primary beam energies.

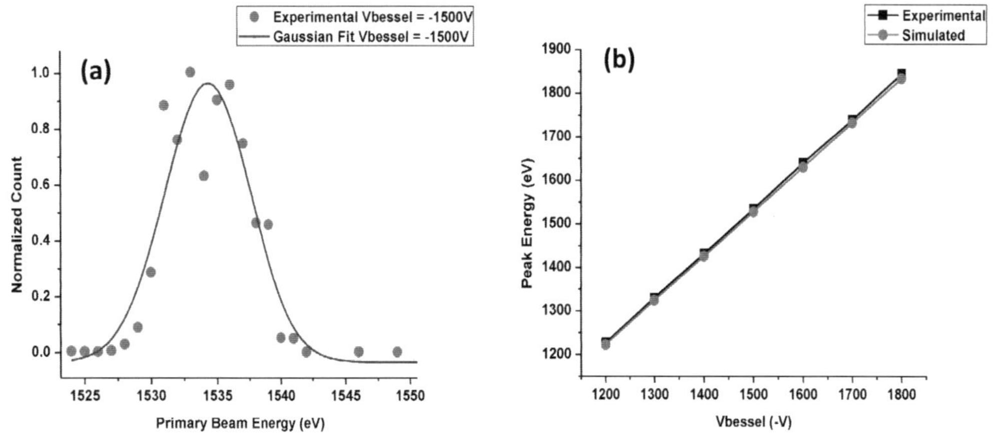

Figure 3. (a) Experimental Normalized Collector Currents (to peak value) for a BB at -1500V. The error in the counts is estimated to be about 10% of the count. An experimental peak energy of 1534eV is observed. (b) Peak Pass energy can be tuned linearly varying the Bessel voltage.

SIMION simulations were also performed for the described setup. Figure 3(a) demonstrates the band-pass action. A peak energy of 1534eV was obtained experimentally with a resolution less than 1%. The error in the count for each point is estimated to be about 10%. An offset between the BB energy (1500 eV) and the peak energy (1534 eV) is due to a convolution of the electron energy and the incident angle demonstrated in figure 2(c). Furthermore, the peak pass energy can be controlled by varying the voltages on the Bessel and output plates. A linear relationship between the peak pass energy and the applied Bessel voltage can be seen in experimental and simulated results of figure 3(b).

4. Conclusion

The simulation and preliminary results of a new SEM electron detector based on a Bessel box (BB) electron energy analyser is presented. The detector is of the band-pass configuration and shows an energy resolution of less than 1%, which should prove useful detecting the energy and angular distribution of secondary electrons. Further work is underway where an array of such BBs is being investigated. The small size of the BB makes it an attractive detector to add-on existing SEM and with better signal-to-noise ratio than the conventional ET detector.

References

[1] Goldstein J. 2003, *Scanning Electron Microscopy and X-ray Microanalysis* (New York: Springer Third edition)
[2] Everhart TE and Thornley RFM 1960 J. of Sci. Inst. **37** 246
[3] Schiwietz G, Beye, M, Kuhn D and Xio G 2015 J. Elec. Spect. Relat. Phenom. **203** 51
[4] Dahl D 2000 Int. J. Mass Spectrom. **200** 3
[5] Allen JD, Durham JD, George KS and Deeds WE 1976 J. Elec. Spect. Relat. Phenom. **8**, 4410.
[6] Stalder RE, Boumsellek S, Zandt TRV, Kenny TR, Hecht MH and Grunthaner FE 1994 J. Vac. Sci. Technol., A **12** 2554

Acknowledgments

The authors would like to acknowledge the support of SIMDALEE2 consortium, FP7-PEOPLE-2013-ITN (Grant agreement No. 606988).

IOP Conf. Series: Journal of Physics: Conf. Series **902** (2017) 012018 doi:10.1088/1742-6596/902/1/012018

Automated quantification of morphology and chemistry from STEM data of individual nanoparticles

Y C Wang[1], T J A Slater[1], T S Rodrigues[2], P H C Camargo[2] and S J Haigh[1*]

[1] School of Materials, University of Manchester, Manchester, M13 9PL, UK
[2] Departamento de Quimica Fundamental, Instituto de Quimica, Universidade de São Paulo, São Paulo, Brazil

*Sarah.haigh@manchester.ac.uk

Abstract. Two novel automated methods for the determination of surface roughness and chemical distribution in individual nanoparticles are presented and applied to nanoparticles synthesised by the galvanic replacement reaction. The two methodologies apply the determination of circumferential and radial line profiles to determine surface roughness and elemental distribution respectively. The surface roughness analysis provides details on localised changes in roughness in comparison to single measures of circularity. The X-ray spectroscopic concentric ring scan outperforms conventional line scans when elemental distribution approximately possesses a spherical symmetry.

1. Introduction

The morphology and elemental distribution of nanoparticles play vital roles in determining their properties, in applications such as catalysis and drug delivery [1, 2]. Specifically, catalytic performance is strongly dependent on surface area and surface roughness. For example, Rodrigues et al. demonstrated distinctive catalytic activity changes for different surface roughness in both nanotubes and nanoparticles [3]. However, studies linking nanoparticle morphology to catalytic property changes are often solely descriptive, with the presentation of a few qualitative electron microscopy images. The lack of quantification of such morphological features leads to difficulties in distinguishing subtle roughness changes between nanoparticle populations.

In addition, energy dispersive X-ray spectrum (EDS) imaging or line scanning are commonly used to provide an overview of the elemental distribution of bimetallic nanoparticles [4–6]. However, low EDS signal counts due to low beam currents or limited acquisition time often result in noisy EDS maps and noisy line scan profiles, which can be inconclusive as a means to explain spatial composition variations.

Herein, we report two techniques based on image processing that aim to provide quantitative surface roughness data and improve signal to noise ratio for EDS elemental distribution analysis. We show that the surface roughness of a nanoparticle can be quantified using a boundary profile extracted from an image acquired by scanning transmission electron microscope high-angle annular dark-field (STEM-HAADF) imaging. We also demonstrate that applying a concentric ring scan to EDS spectrum image data for spherical nanoparticles presents distinct advantages compared to simple one dimensional (1D) line scans when determining the elemental surface segregation of a bimetallic nanoparticle.

2. Experimental Methods

The chosen nanoparticles of AgPt and AgPd in this study have shown high efficiency as catalysts in the conversion of 4-nitrophenol. They were synthesised by the galvanic replacement reaction between either $PtCl_6^{2-}$ or $PdCl_4^{2-}$ in aqueous solution and Ag nanosphere templates, using polyvinylpyrrolidone (PVP) as a stabilizer. The detailed synthesis route was reported in references [3], [7]. The synthesised nanoparticle solutions were then dropped on to a Cu grid layered with a holey carbon film.

Imaging was performed using an FEI Titan G2 80-200 S/TEM, which is equipped with an (X-FEG) high brightness source, probe-side aberration correction, and the ChemiSTEM™ Super-X EDS detector that consists of four silicon drift detectors (SDDs). EDS spectrum imaging was performed with a probe current of 180 pA, 512 x 512 pixel areas, a pixel size of 0.1 nm, a dwell time of 40 μs and a total acquisition time of 38 minutes. After data acquisition, image processing was applied to STEM-HAADF images of multiple particles to separate individual nanoparticles from the background. The image processing steps consisted of image background subtraction by a rolling ball algorithm to eliminate uneven intensity distributions, thresholding to produce a binary image mask, watershed segmentation to isolate multiple touching nanoparticles and, finally, replacing each masked isolated nanoparticle into a zero-intensity background box of 501x501 pixels (pixel size 0.1 nm) with the box centred on the nanoparticle centre of geometry.

Surface roughness analysis was then performed using the following procedure: 1) delineate nanoparticle boundaries; 2) calculate distances from the nanoparticle centre to the nanoparticle boundary for the whole circumference; 3) plot the distance profile against angle from 0° (12 o'clock direction) to 360° as shown in Figure 1 a and b; 4) at 30° intervals, measure the difference between the maximum and minimum distance within each interval; 5) calculate the median difference for the 12 intervals to provide a quantitative description of the surface roughness of each nanoparticle.

For the EDS concentric ring scan, the nanoparticle was firstly segmented and placed in a box as the procedures mentioned above by using the simultaneously acquired HAADF image. For every pixel on the isolated nanoparticle, the pixel coordinates at raw and boxed HAADF images were known. Then each EDS spectra associated with pixels were copied from raw EDS image to boxed image to generate the EDS spectrum image of the segmented nanoparticle. Afterwards, EDS counts on each concentric ring were summed and normalised by the number of pixels on the respective concentric ring. The final plot of normalised EDS counts on each concentric ring against radial distance from particle centre to the boundary were then depicted, as shown in Figure 2d.

3. Results and Discussion

3.1. Surface roughness analysis

For a convex object such as a three dimensional (3D) nanoparticle, the boundary of its 2D projection is roughly equivalent to the 1D representation of its 2D surface roughness. Therefore, by quantifying the fluctuation of the boundary profile, the nanoparticle's surface roughness can be measured. As there is no surface fluctuation for a perfect sphere, i.e. every distance from the surface to the sphere centre remain the same, the 2D projection and corresponding boundary profile of a sphere would be that displayed in Figure 1a and b, and the surface roughness would be zero.

The real nanoparticle, as shown in Figure 1c, however, presents large variations in its boundary profile (Figure 1d). These variations can be attributed to three main causes: 1) variations due to a rough surface, 2) shape variations (i.e. the difference between a circle and an oval that have smoothly varying surfaces but different measures of circularity) and 3) size variations, larger particle sizes could result in larger variations in distance values. In order to differentiate the surface roughness variations from the other two factors, we divide the profile into angular intervals. The size of the intervals is chosen such that the boundary segment is not so large that the maximum, minimum and variations measured within each interval are a result of the particle shape variation. In addition, the interval size should not be too small that it is only measuring noise (experimental error) in the data. For our data 12 angular intervals (each 30°) was found to be a suitable compromise condition. Within each interval, we measure the difference between the maximum and minimum distances between centre and surface. The final value to represent the surface roughness could be the median or mean difference calculated from all intervals. For this nanoparticle, the general peak-to-valley length for each segment of the nanoparticle projected surface is either 2 nm (median) or 1.6 nm (mean). The circularity, calculated by $4\pi \times \frac{[\text{Area}]}{[\text{Perimeter}]^2}$, cannot straightforwardly describe the surface roughness in this manner, and is prone to greater variations with varying geometric shape of the nanoparticles. The surface roughness could also be represented by the standard deviation of differences in each interval. If a particle is not a nearly circular shape in 2D projection but a triangular or a square, the boundary profile could be compensated by the variation of an equivalent shape. Then the boundary profile is similar to an representation of a flat surface roughness and can be measured by flat surfaces roughness standard ISO 4287:1997 using the extracted surface profile.

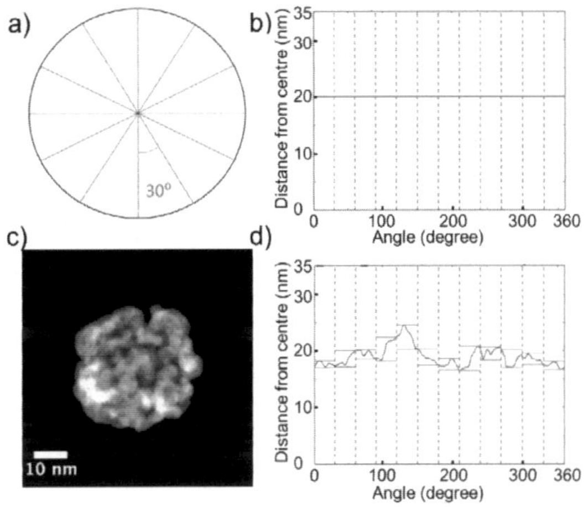

Figure 1. Surface roughness analysis for a AgPt nanoparticle synthesised by galvanic replacement. (a) Schematic illustrating measurement for the centre to boundary distances for the projection of a perfect sphere. (b) Boundary profile calculated for the circle depicted in (a). (c) Real STEM-HAADF projection image of a nanoparticle. (d) Boundary profile calculated for the image in (c).

3.2. EDS concentric ring scan

The raw EDS map of an AgPd nanoparticle without any binning or smoothing is shown in Figure 2a. The total acquisition time was 38 min in order to maximum EDS counts on the spectrum image. The nanoparticle was monitored continuously during spectrum imaging and STEM-HAADF image remained nearly same to the original. From the unprocessed EDS elemental map, the particle appears to possess a shell structure with a hollow interior. However, the detailed locations of elements within the shell is unclear. Typically, we apply a 1D line scan across the nanoparticle, at the position indicated by the red box on Figure 2a to assess the extent of compositional segregation. The measured EDS line profile is presented in Figure 2b. The width of the line is set to 50 pixels, equal to 5 nm to maximise counts per unit length. The line width is approximately one third of the nanoparticle diameter and the scanned area (red box area) is 10% of the whole image. The poor signal to noise ratio of the resulting line scan profile yields limited information on elemental distribution in the particle. The reasons for the noisy spectral data are a combination of small EDS collection solid angle (~0.7 srad), limited acquisition time and low X-ray cross sections for both Ag K lines and Pd K lines (Ag and Pd L lines are overlapping).

EDS concentric ring scan provides a means to use a much greater proportion of the acquired EDS counts than the 1D line scan. The schematic of the concentric ring scan is illustrated in Figure 2c and the resultant EDS profile is shown in Figure 2d. The EDS intensities of each ring are normalised by the number of pixels in the circumference at each point. At the centre of the particle (less than 10% radius), all pixels are calculated, producing the flat line at the start of each EDS profile in Figure 2d. The low intensity of EDS profiles in the centre is due to the particle's hollow interior and the peaks at ~13 nm from the centre demonstrate the shell structure. Figure 2d displays a Ag single peak, which suggests the middle of the shell of this nanoparticle is of high silver content. Palladium, in contrast, gives two peaks on the left and right side of the Ag peak suggesting Pd segregates to the outer and inner surfaces of the shell.

Although this study uses a sample with a near circular 2D projection, the idea of the concentric ring scan can be expanded to any projected shapes in which a radial elemental distribution can be found. It just requires changing the EDS counts calculation of each ring to the boundary of the projection, if we can obtain the boundary profile using the aforementioned method.

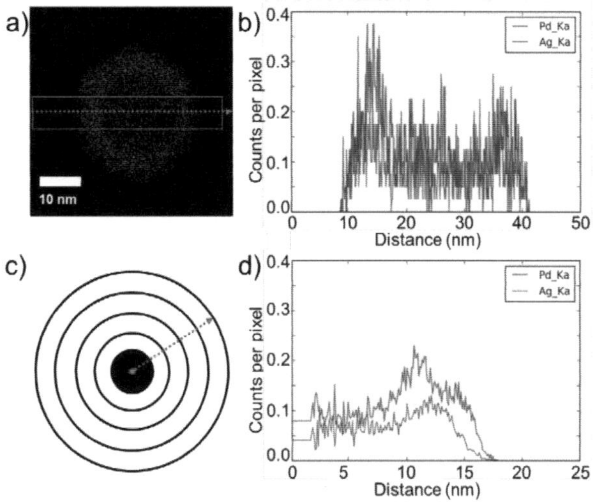

Figure 2. (a) EDS raw map displaying X-ray counts of Ag (green) and Pd (magenta) for a AgPd nanoparticle synthesised by galvanic replacement. (b) 1D EDS line scan profile at the position of the red rectangular and the direction of the dashed arrow on (a). (c) Schematic of the concentric ring scan. (d) EDS concentric ring scan profile at the direction of the red dashed arrow on (c) for the particle shown in (a). Both scans have been normalised by the number of pixels per unit length.

4. Conclusions

We have presented two automated methodologies to quantify surface roughness and compositional variations for spherical nanoparticles. The surface roughness analysis is useful to quantify nanoparticle populations in an automated manner and thus provide statistical conclusions to differentiate subtle surface roughness variations. The EDS concentric ring scan as an alternative to a line scan, yielding higher signal to noise data where the specimen fulfils the spherical symmetry requirement.

Acknowledgements

Y.C.W thanks the PhD funding of China Scholarship Council (CSC) and University of Manchester. The authors wish to acknowledge the support from HM Government (UK) for the provision of the funds for the FEI Titan G2 80-200 S/TEM as well as the EPSRC and ERC for financial support.

References

[1] Wang X, Feng J, Bai Y, Zhang Q, and Yin Y 2016 *Chem. Rev.* **116** 10983–1060
[2] Laramy C R, Brown K A, O'Brien M N and Mirkin C A 2015 *ACS nano* **9** 12488-95
[3] Rodrigues T S *et al.* 2016 *Catal. Sci. Technol.* **6** 2162–70
[4] Lewis E A *et al.* 2014 *Nanoscale* **6** 13598-605
[5] Slater T J A *et al.* 2014 *Nano Lett.* **14** 1921-6
[6] Lewis E A *et al.* 2014 *Chem. Commun.* **50** 10019-22
[7] da Silva A G M *et al.* 2017 *Chem. Commun.* **50** DOI 10.1039/c7cc02352a

High quality lamella preparation of gallium nitride compound semiconductor using Triple Beam^{TM} system

T Sato[1, 3], K Nakano[1], H Matsumoto[1], S Torikawa[2], I Nakatani[2], M Kiyohara[2] and T Isshiki[3]
[1] Hitachi High-Technologies Corp., 1040, Ichige, Hitachinaka-shi, Ibaraki, 312-0058, Japan
[2] Hitachi High-Tech Science Corp., 36-1, Takenoshita, Oyama-cho, Shizuoka, 410-1393 Japan
[3] Kyoto Institute of Technology, 1, Matugasaki-hashikamicyo, Sakyo-ku, Kyoto, 606-8585, Japan

E-mail: takahiro.sato.fv@hitachi-hightech.com

Abstract. Gallium nitride (GaN) compound semiconductors have been known to be very sensitive to Ga focused ion beam (FIB) processing. Due to the nature of GaN based materials it is often difficult to produce damage-free lamellae, therefore applying the Triple Beam^{TM} system which incorporates an enhanced method for amorphous removal is presented to make a high quality lamella. The damage or distortion layer thickness of GaN single crystal prepared with 30 kV Ga FIB and 1 kV Ga FIB were about 17 nm and 1.5 nm respectively. The crystallinity at the uppermost surface remained unaffected when the condition of 1 kV Ar ion milling with the Triple Beam^{TM} system was used. The technique of combining traditional Ga FIB processing with an enhanced method for amorphous layer removal by low energy Ar ion milling allows us to analyse the InGaN/GaN interface using aberration corrected scanning transmission electron microscopy at atomic resolution levels.

1. Introduction

The GaN compound semiconductor is a significant candidate material for light emitted diode circuits or power electronics devices due to its excellent physical properties [1-2]. The electrical performance is strongly related to the nano structure, as a result, the need for analysis of compound semiconductors at the atomic scale using transmission electron microscope (TEM) or scanning transmission electron microscope (STEM) has been of increasing interest. However, it is known that the GaN based compound semiconductor is especially sensitive to the Ga ion of FIB during the lamella prodcution process. A damage layer of several tens of nano meter thickness is formed, degrading the image quality. We have developed the Triple Beam^{TM} system to reduce Ga ion milling related damage. “Triple Beam” is a trademark of Hitachi High-Tech Science Corporation. This technology has been applied to lamella preparation for several types of material such as silicon semiconductor integrated circuits or ceramics. In this study, low damage and high quality lamella preparation condition of GaN compound semiconductor is discussed, and evaluation of the analytical effectivity along the interface at atomic scale resolution.

IOP Conf. Series: Journal of Physics: Conf. Series **902** (2017) 012019 doi:10.1088/1742-6596/902/1/012019

2. Experimental

The Hitachi NX2000 FIB-SEM Triple Beam™ system was used for sample preparation [3]. Figure 1 shows the diagram of Triple Beam™ system. A FIB column, scanning electron microscope (SEM) column, and Ar/Xe column are installed on the same specimen chamber, designed with each beam to coincide at the same point. The enhanced removal of FIB induced damage using low energy Ar or Xe ion milling is possible in the NX2000 system. A bulk lamella was extracted from the bulk sample using the patented Hitachi micro-sampling in-situ pick up system inside the specimen chamber of the NX2000 system [4].

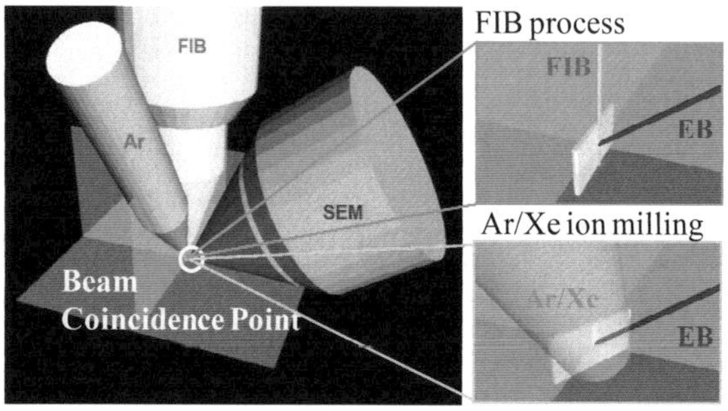

Figure 1. The diagram of Triple Beam™ system.

In the first experiment, the damage layer thicknesses of FIB and ion milling on a GaN single crystal was measured. The damage layer of Ga FIB was formed by acceleration voltage of 30 kV, 10 kV, 5 kV, 1 kV, and finally 1 kV for Ar ion milling. The top surface of the damage layer was protected using electron beam deposition. The five lamellas were prepared to measure each damage layer. An Hitachi HF-3300 cold field emission (FE)-TEM operated at 300 kV was used for the measurement and the evaluation of crystallinity. Optimum lamella preparation condition was determined by those measurement results. In the second experiment, the low damage lamella of nitride semiconductor substrate with InGaN/GaN multi quantum well (MQW) was prepared with optimized sample preparation conditions [5]. The STEM image quality of these lamella was compared. Especially, the sharpness of InGaN/GaN interface was evaluated using an Hitachi HD-2700 aberration corrected STEM and Oxford Instruments X-Max^{N}100TLE dual silicon drift detector (SDD) system for energy dispersive X-ray (EDX) analysis. The 100 mm^2 window size and window less detector was adopted for high sensitivity EDX analysis.

3. Results and discussion

Figure 2 shows the high resolution TEM images of GaN single crystal with a formed damage layer intentionally produced. Images showing the results of FIB 30 kV (a), FIB 10 kV (b), FIB 5 kV(c), FIB 1 kV (d) and Ar 1 kV (e), respectively. The electron beam condition for the images is on the <1$\bar{1}$00> axial illumination. The diffraction contrast, within the dark area in the TEM images, relates with crystal distortion which was observed in FIB preparation lamella. The thickness of the layer is about 17 nm shown in figure 2(a), which decreased with the decline of FIB acceleration voltage. By contrast, the dark area was removed, revealing the crystallinity of GaN which was kept at the top surface in figure 2(e). It is found that Ar ion milling is effective in removing the FIB damage layer.

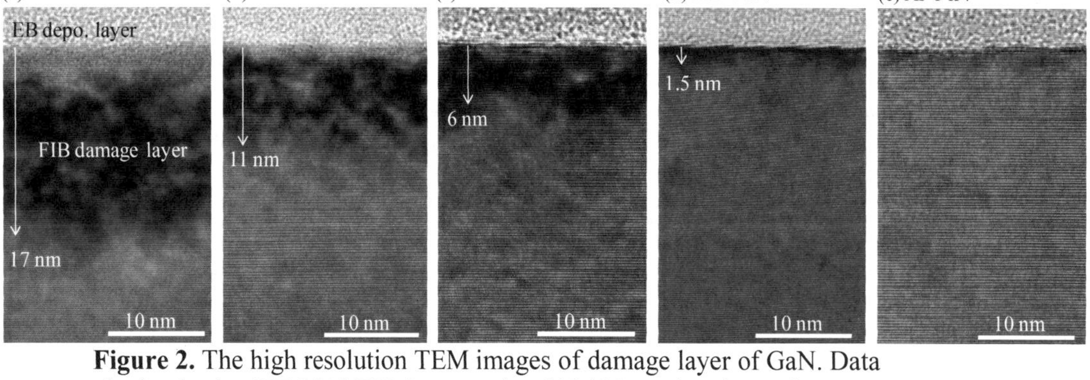

Figure 2. The high resolution TEM images of damage layer of GaN. Data obtained using HF-3300 TEM operated at 300 kV acceleration voltage.

Figure 3 shows the TEM observation result of InGaN/GaN MQW lamella. Images showing the results of FIB 30 kV (a), FIB 5 kV(b), FIB 2 kV (c), Ar 0.5 kV (d) and Xe 1 kV (e), respectively. The incident angle of Ar 0.5 kV is a 20 degree, and Xe 1 kV is a 5 degree. The electron beam incident direction is on the <11$\bar{2}$0> axial illumination. The light and shade of TEM image of figure 3(a) is damage, it degrades the image quality. It was reduced with the decline of FIB acceleration voltage, but it was not enough even FIB 2 kV. The results of Ar and Xe milling show clear and smooth TEM image without distortion contrasts, and five MQW layers are clearly observed with high contrast difference. TEM image of Xe milling is smoother than Ar milling. It might be considered that the damage depth of Xe is shallower than Ar, because the mass number of Xe is heavier than Ar. But, due to the difference in milling conditions, further comments are not possible.

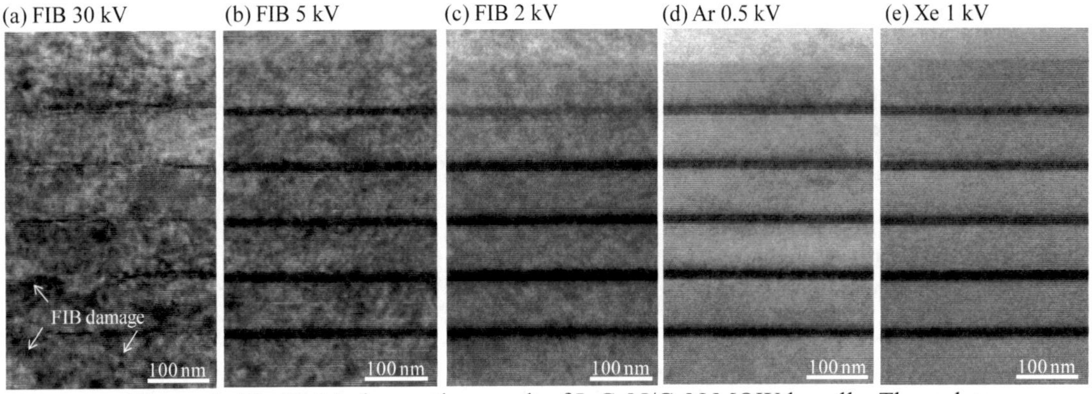

Figure 3. The TEM observation result of InGaN/GaN MQW lamella. These data were taken by using HF-3300 TEM operated at 300 kV acceleration voltage.

Figure 4 shows high resolution annular dark field (ADF)-STEM images of InGan/GaN MQW lamella processed with Ar 1 kV milling. In figure 4 (a), three layers of InGaN were observed with higher contrast due to the difference of atomic number with GaN. In figure 4 (b), an InGaN layer consisted of eight or nine atomic layers, lower interface of InGaN/GaN is very sharp, but upper interface of GaN/InGaN is rough. The crystal growth condition of the InGaN/GaN MQW contributes to the atomic arrangement of interface, high quality lamella can distinguish this difference.

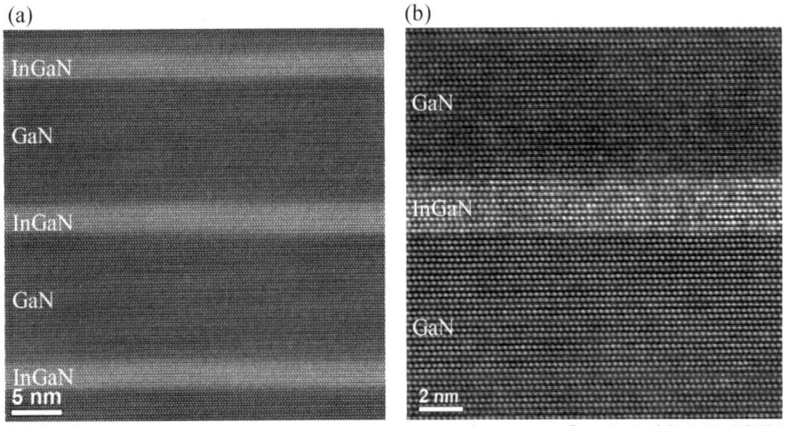

Figure 4. The high resolution ADF-STEM images of InGaN/GaN MQW lamella. 200 kV HD-2700 was used for observation.

Figure 5 shows atomic resolution EDX mapping and line analysis result. Atomic column site of Ga or In were visualized. Line analysis was extracted from EDX spectrum imaging data. It is confirmed that the InGaN layer consists of eight or nine atomic layers. Distribution of In atom is not uniform in map image, it should be considered that the In atom fluctuation in InGaN layer was detected at atomic resolution scale [6].

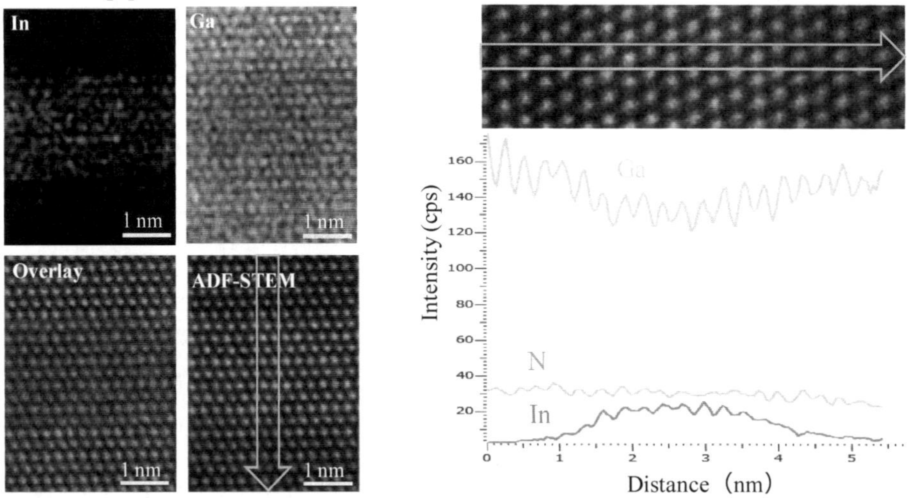

Figure 5. Atomic resolution EDX mapping and line analysis. HD-2700 equipped with X-Max^N 100TLE SDD was used for EDX analysis. Pixel size is 128 ×100, acquisition time is about ten minute.

4. Conclusions

We have developed the Triple Beam^{TM} system to remove the Ga induced FIB damage layer and applied it to high quality lamella preparation of GaN compound semiconductor. As a result, the damage layer formed by FIB was perfectly removed by Ar or Xe ion milling, and the TEM image quality of InGaN/GaN MQW lamella was improved, these prepared lamellae allow the analysis at atomic scale resolution levels. It was confirmed that the Triple Beam^{TM} system of NX2000 is a powerful tool to remove Ga induced FIB damage, and it enables the preparation of low damage and high quality lamella of GaN compound semiconductors or other materials.

References

[1] Amano H, Sawaki N, Akasaki I and Toyoda Y 1986 *Appl. Phys. Lett.* **48** 353-355
[2] Ohno Y, Kuzuhara M, 2001 *IEEE Trans. on Electron Devices* **48** 517-523
[3] Takahashi H *et al*, 2006 *Proc. 16th International microscopy congress* p 1104
[4] Ohnishi T *et al*, 1999 *Proc. 25th int'l Symposium for Testing and Failure Analysis* p 499
[5] Kaga M, Morita T, Kuwano Y, Yamashita K, Yagi K, Iwaya M, Takeuchi T, Kamiyama S and Akasaki I 2013 *Japanease Jounal of Applied Physics* **52**(8) 5208JH07
[6] Sugahara T, Sato H, Hao M, Naoi Y, Kurai S, Tottori S, Yamashita K, Nishino K, Linda T. Romano and Sakai S 1998 *Japanease Jounal of Applied Physics* **37** L1195

Acknowledgements

We would like to thank Professor Satoshi Kamiyama from Meijo University who provided the InGaN/GaN nitride semiconductor and Jamil J. Clarke at Hitachi High-Technologies Corporation in America for his contribution to support this effort.

IOP Conf. Series: Journal of Physics: Conf. Series **902** (2017) 012020 doi:10.1088/1742-6596/902/1/012020

Evaluating focused ion beam patterning for position-controlled nanowire growth using computer vision

A B Mosberg[1], S Myklebost[1], D Ren[2], H Weman[2], B O Fimland[2] and A T J van Helvoort[1]

[1] Department of Physics & [2] Department of Electronic Systems, Norwegian University of Science and Technology (NTNU), 7491 Trondheim, Norway

E-mail: aleksander.b.mosberg@ntnu.no

Abstract. To efficiently evaluate the novel approach of focused ion beam (FIB) direct patterning of substrates for nanowire growth, a reference matrix of hole arrays has been used to study the effect of ion fluence and hole diameter on nanowire growth. Self-catalyzed GaAsSb nanowires were grown using molecular beam epitaxy and studied by scanning electron microscopy (SEM). To ensure an objective analysis, SEM images were analyzed with computer vision to automatically identify nanowires and characterize each array. It is shown that FIB milling parameters can be used to control the nanowire growth. Lower ion fluence and smaller diameter holes result in a higher yield (up to 83 %) of single vertical nanowires, while higher fluence and hole diameter exhibit a regime of multiple nanowires. The catalyst size distribution and placement uniformity of vertical nanowires is best for low-value parameter combinations, indicating how to improve the FIB parameters for positioned-controlled nanowire growth.

1. Introduction

III-V semiconductor nanowires are a promising material system for the creation of future optoelectronic devices [1]. Using lithography-based patterning of an oxide mask, well-defined nucleation sites are placed at will, often in homogeneous patterns to ensure identical growth conditions. This approach has been successful in achieving large arrays of similar nanowires, but wire-to-wire variations still need to be evaluated [2]. In order to improve nucleation and further reduce variation, direct oxide patterning is expected to be advantageous, allowing for more flexible hole geometry to optimize patterning and nanowire nucleation conditions.

In this work, focused ion beam (FIB) is used to pattern growth substrates for self-catalyzed GaAsSb nanowires grown using molecular beam epitaxy (MBE) [3]. The FIB patterning enables direct patterning of the oxide mask and is more flexible than the lithography based techniques conventionally used. The parameter space to optimize nanowire growth is efficiently explored on a single growth sample. To evaluate the effect of milling conditions on nanowire growth, a sufficient quantity of nanowires need to be characterized in an objective and efficient way. This is especially important for evaluating this rather novel approach to patterning for nanowire growth. By utilizing feature detection techniques from the field of computer vision [4] to automatically detect nanowires from top-down scanning electron microscope (SEM) images, a detailed and objective characterization of the parameter space is achieved.

2. Methods and materials

Self-catalyzed GaAsSb nanowires were grown in a Varian GEN II Modular MBE system [5]. To pattern the substrate for position-controlled nanowire growth, a FEI Helios NanoLab 600 DualBeam FIB was used at 30 kV to mill a growth matrix into a Si(111) wafer with a 40 nm thick SiO_x film (figure 1(a)). The growth matrix consists of 8 × 8 hole arrays with linearly increasing combinations of ion fluence (0.418 – 3.329 10^{17} ions/cm^2) and hole diameter (10 – 80 nm). Each hole array contains 270 holes in a hexagonal pattern with 1 µm pitch (figure 1(b)). The sample was cleaned using 1 % HF for 150 s before insertion in the MBE system.

Each array was imaged with 5 kV SEM in the DualBeam (figure 1(c)). Top-down images were then used as input for feature detection, implemented in open source Python libraries. By optimizing the SEM contrast for computer vision, Ga catalyst droplets were identified and used to count and characterize nanowires (figure 2(a)). A lattice based on the FIB-milled pattern, fitted to the detected nanowires, assigns each nanowire to a lattice point corresponding to a FIB-milled hole. The Python code for detection and analysis has been made freely available [6].

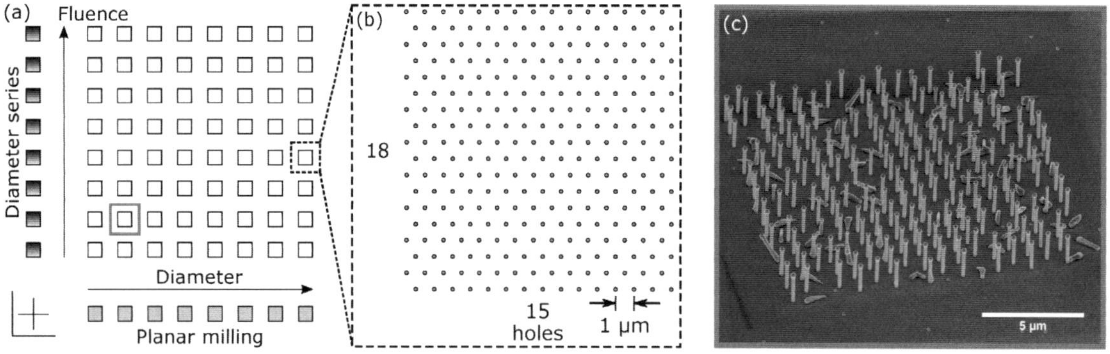

Figure 1. (a) Reference design created in FIB: ion fluence - diameter matrix with 64 arrays and supplemental reference fields. (b) Each array consists of 15 × 18 holes. (c) Tilted SEM of an array after growth, with a high yield of single vertical nanowires.

3. Results and discussion

Vertical nanowire growth is observed in all milled arrays, demonstrating the viability of FIB as an alternative patterning technique for nanowire growth substrates. There is a general trend that smaller, shallower (i.e., lower fluence) holes give a high yield of single vertical nanowires per hole (figure 2(a,b)). For wider and deeper holes multiple vertical nanowires are observed per hole (figure 2(d,e)). Between the single and multiple wire regimes, total nanowire yield is lower, dominated by more parasitic growth. This variety demonstrates the necessity of FIB patterning optimization to obtain full growth control among a broad range of possible structures.

The feature detection is consistently able to detect Ga droplets and distinguish them from other features in the SEM image, such as stray Ga droplets on the sample surface (figure 2(a,b)). This allows for automatic detection of nanowires giving a quantitative and objective analysis of how the FIB milling parameter space affects nanowire growth. The maximal yield of single nanowires, 83 %, was observed for two arrays at lower parameter combinations (figure 2(c)). At higher fluence and diameter, array have more variance with several holes containing one or multiple nanowires. The computer vision-based approach ensures correct characterization and classification of growth regimes. For example, a maximal yield of 35 % for two nanowires per hole is found (figure 2(d-f)) in the same array where one of the lowest yields of single nanowires (30 %) is observed. For larger arrays of nanowires, the convenience and reliability of automated over manual characterization becomes more important and ensures reproducibility.

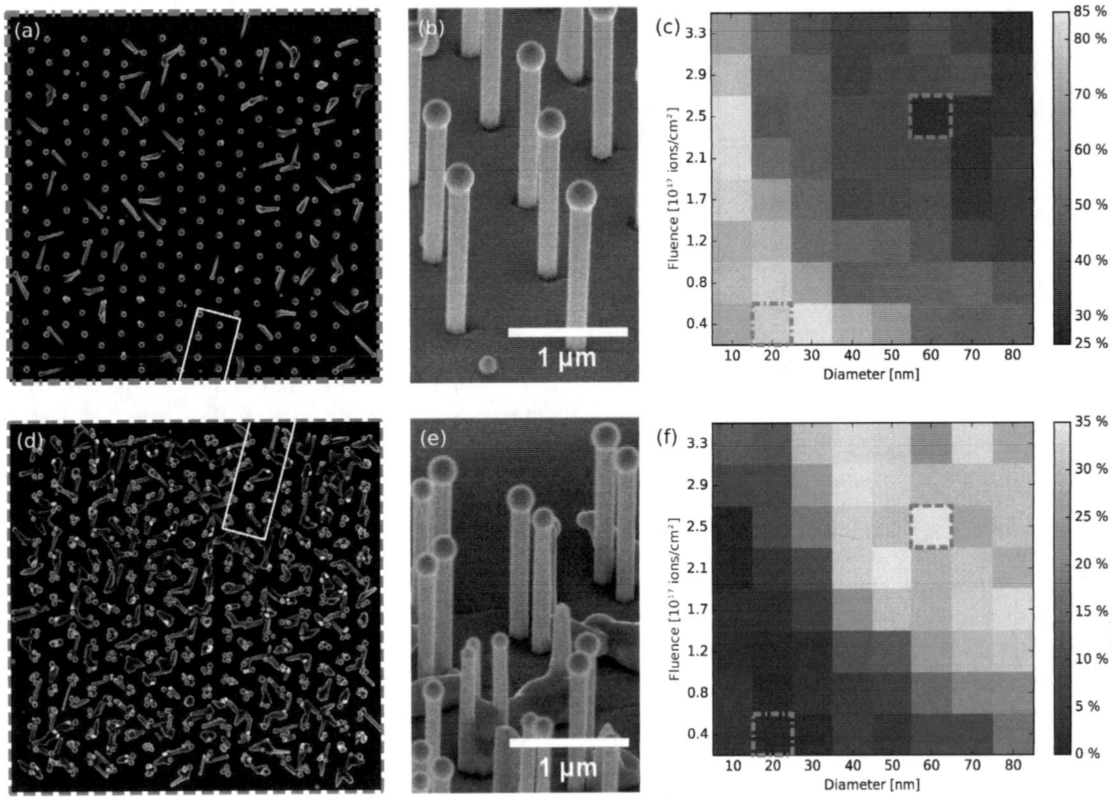

Figure 2. (a) Detected droplets overlaid on SEM image for low parameter combination array. (b) Tilted-view SEM image of single nanowire regime, from frame in (a). (c) Single nanowire yield plot across the growth matrix. (d) Detected droplets on SEM image for higher parameter combination. (e) Tilted-view SEM image of multiple-nanowire regime, from frame in (d). (f) Yield plot for two nanowires per hole across the growth matrix.

Taking advantage of the breadth of information provided by computer vision on SEM images, mean droplet diameter and displacement from the fitted lattice positions can be evaluated (figure 3(a,b)). The catalyst diameter decreases with increasing ion fluence and hole diameter. This trend can be explained by the additional parasitic growth and multiple nanowires per hole for higher fluence-diameter combinations (figure 2(d)). With constant Ga flux across the sample during MBE growth, the Ga supply per wire decreases with increasing number of droplets.

At the same time, the mean deviation from fitted lattice positions increases with both fluence and diameter (figure 3(b)). Plotting the deviations in scatter plots (figure 3(c), shown for the growth matrix extremes), two distinct effects are identified: First, higher diameter holes consistently have nanowires nucleating further out from lattice centers, indicating that nanowires seem to preferentially nucleate along the hole side walls rather than in the hole center. For wider holes, the increased circumference allows for multiple nucleation sites. Second, deeper (i.e., higher fluence) holes lead to more off-center nucleating nanowires (figure 3(c(i))) and less radially symmetric displacement. The off-center cluster of nanowire nucleation is believed to be linked to slow ion beam blanking, resulting in an asymmetric hole edge (visible in figure 2(b), blanking lines visible in figure 2(e)) leading to an uneven distribution of nanowire nucleation sites within a single hole. This can be remedied by the use of a faster beam blanker or alternative scan strategies when deeper holes are desired.

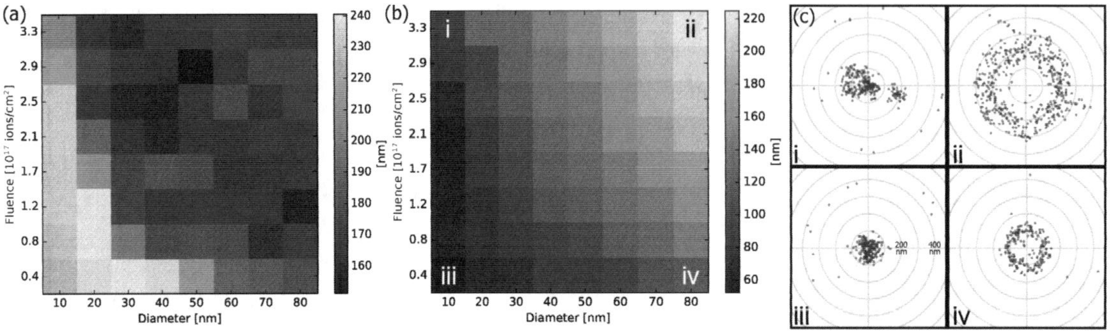

Figure 3. (a) Median droplet diameter per array. (b) Median droplet distance from fitted lattice position per array. (c) Scatter plots of droplet displacements from fitted lattice, for the four arrays indicated in (b).

As droplet size and contact angle has been shown to influence both crystallinity and composition [7], the variation in yield, droplet size, and effective Ga/As ratio across the growth matrix is expected to influence the GaAsSb nanowire composition and optoelectronic properties [5]. To further investigate this, micro-photoluminescence spectroscopy, electrical probing of single nanowires, and transmission electron microscopy of the nanowire-substrate interface should be performed for the different arrays and correlated to the results from computer vision-based studies. In this way the trends in growth results can be linked to FIB patterning parameters to systematically study and achieve the optimal properties for nanowire-based devices.

4. Conclusion

FIB milling has been systematically studied as a promising and flexible direct patterning method for self-catalyzed nanowire growth substrates. Computer vision was successfully applied to detect Ga droplets on top of vertical nanowires for a substantial number of holes (17280 holes across 64 different arrays) on a single sample and thereby shed light on how the FIB milling parameters affect nanowire growth. The ion fluence and hole diameter were found to affect vertical nanowire yield, number of nanowires per hole, droplet size distribution, and nanowire displacement from patterned lattice position. These nanowire characteristics can thus be correlated and optimized in future growth trials.

Acknowledgments

The authors acknowledge NTNU and the Research Council of Norway for financial support via the initiative Enabling technologies: Nanotechnology, NORFAB (grant 197411), NORTEM (grant 197405), and the FRINATEK-program (grant 214235).

References

[1] Joyce H J *et al.* 2011 *Progress in Quantum Electronics* **35** 23–75
[2] Nilsen J S, Reinertsen J F, Mosberg A, Fauske V T, Munshi A M, Dheeraj D L, Fimland B O, Weman H and van Helvoort A T J 2015 *Journal of Physics: Conference Series* **644** 012007
[3] Detz H, Kriz M, Lancaster S, MacFarland D, Schinnerl M, Zederbauer T, Andrews A M, Schrenk W and Strasser G 2017 *Journal of Vacuum Science & Technology* B **35** 011803
[4] Lindeberg T 1998 *International Journal of Computer Vision* **30** 79–116
[5] Ren D *et al.* 2016 *Nano Letters* **16** 1201–9
[6] NWstats GitHub repository URL `https://www.github.com/NWstats/NWstats`
[7] Munshi A M, Dheeraj D L, Todorovic J, van Helvoort A T J, Weman H and Fimland B O 2013 *Journal of Crystal Growth* **372** 163–9

IOP Conf. Series: Journal of Physics: Conf. Series **902** (2017) 012021 doi:10.1088/1742-6596/902/1/012021

Strategy for reliable strain measurement in InAs/GaAs materials from high-resolution Z-contrast STEM images

Maryam Vatanparast[1], Per Erik Vullum[1, 2], Magnus Nord[3], Jian-Min Zuo[4], Turid W. Reenaas[1] and Randi Holmestad[1]

[1] Department of Physics, Norwegian University of Science and Technology - NTNU, 7491, Trondheim, Norway

[2] SINTEF Materials and Chemistry, 7465, Trondheim, Norway

[3] SUPA, School of Physics and Astronomy, University of Glasgow, Glasgow, United Kingdom

[4] Dept. of Materials Science and Engineering, University of Illinois, Urbana, 61801, USA

Email: maryam.vatanparast@ntnu.no

Abstract. Geometric phase analysis (GPA), a fast and simple Fourier space method for strain analysis, can give useful information on accumulated strain and defect propagation in multiple layers of semiconductors, including quantum dot materials. In this work, GPA has been applied to high resolution Z-contrast scanning transmission electron microscopy (STEM) images. Strain maps determined from different g vectors of these images are compared to each other, in order to analyze and assess the GPA technique in terms of accuracy. The SmartAlign tool has been used to improve the STEM image quality getting more reliable results. Strain maps from template matching as a real space approach are compared with strain maps from GPA, and it is discussed that a real space analysis is a better approach than GPA for aberration corrected STEM images.

1. Introduction

GPA is a novel and widely applied technique that can be used for quantitative measurements of lattice strains in both high-resolution transmission electron microscopy (HRTEM) and atomic resolution high angular annular dark field (HAADF) STEM images. The method is based on the assumption that there exists a constant spatial relationship between the intensity maxima and the location of lattice planes in the studied material [1]. This relationship appears in the form of a spatial shift of the intensity maxima positions with respect to the lattice. With the development of probe aberration correctors in STEM, sub-Ångström resolution now can be achieved with relatively uniform contrast, providing precise atomic-column positions and clear contrast to map the strain field. Nevertheless, additional precautions must be taken in experimental images to avoid unwanted effects from the microscope or the specimen that can produce incorrect strains. Questions also remain on the best strategy for strain mapping, and in GPA also to which two *g* vectors should be chosen in the Fourier transform to create the strain maps [2].

2. Experimental Details

In the present work, the GPA technique is applied to experimental HRSTEM images of InAs/GaAs quantum dot (QD) based intermediate band solar cell (IBSC) materials. The aim is to study strain relaxation mechanisms during the growth process of III–V semiconductors. The IBSC utilizes multiple band gaps in order to increase the efficiency beyond the Shockley-Queisser limit [3]. A necessary but challenging step towards fabricating high efficiency QD based IBSCs is to precisely control the QD structures, including strain, which requires accurate structure information. The InAs/GaAs QD materials studied here were grown by molecular beam epitaxy (MBE), in the Stranski-Krastanov (SK) growth mode on [001] GaAs substrates. The lattice mismatch between two materials in SK growth is used to create the QDs. For pure InAs grown epitaxially on GaAs, the mismatch is 7.2%, and the mismatch is smaller when there is a mixing between In and Ga. Thus, the maximum strain is expected at 7.2%. This strain can be a problem when growing multiple layers of QDs, due to formation of defects, which might act as recombination centers. To compensate for the inherent strain, one can introduce nitrogen in the spacer layers. The small N atoms are supposed to cancel out the strain introduced by the large In atoms. The experimental HAADF-STEM images in Figs. 1 and 4 were taken using a double Cs-corrected, JEOL ARM-200CF microscope, at 200 kV, and the HAADF-STEM image in Fig. 2 was acquired using a 300kV FEI Titan with a Cs probe corrector. All images were taken with the electron beam parallel to the crystallographic [110] direction.

3. Results and discussion

A major issue in using HAADF-STEM images for strain mapping is the scan noise due to the flyback errors in the STEM scanning coils. A strategy to reduce the scan noise, and also to improve the quality of HAADF-STEM images, is to use multiple images recorded with a short scan dwell time and at 0° and 90° scan directions. Fig. 1 compares the effectiveness of this strategy by comparing the ε_{yy} map obtained from a single HAADF-STEM image and the scan noise corrected HAADF-STEM image using the SmartAlign tool [4]. We have used a very short dwell time of 2 μs, and recorded up to 40 frames at 0 and 90° rotations. The SmartAlign tool [4] superposes this image stack into a single image. Much of the noise and horizontal, artificial strain lines seen in the strain map from a single STEM image acquired with the pixel time of 38.8 μs are removed in the strain map of the HAADF-STEM image after applying SmartAlign with 90° rotation between every image in the stack that is shown in Fig.1 (b). We observe in Fig. 1 (d) that the systematic lines are removed from the strain map.

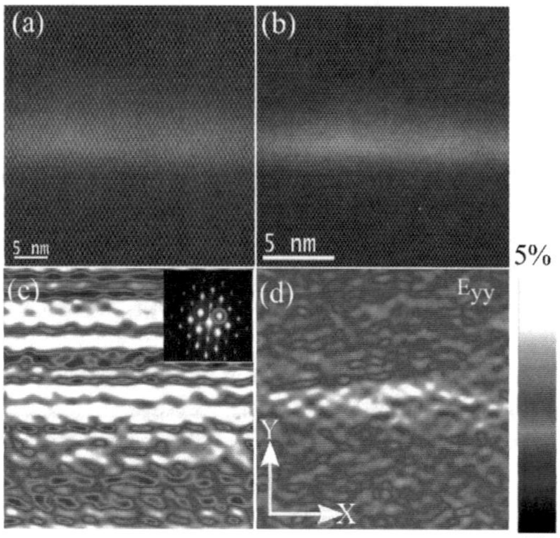

Figure 1. HAADF-STEM image of InAs/GaAs QD, before (a) and and after (b) applying the SmartAlign with 90° rotation, (c) ε_{yy} strain measured from (a), (d) ε_{yy} strain measured from (b). The Bragg spots highlighted in the inset FFT in (c) are used to create strain maps.

The GPA requires two independent Bragg spots as a pair in the Fourier spectrum. The choice of the pair could have a large impact on the GPA results. In Fig. 3, GPA has been applied on different pairs in the fast Fourier transform (FFT) of the single HAADF-STEM image in Fig. 2. Strain maps obtained from (004)/(220), (002)/(220) and {111} pairs (a-c) using GPA show that in case of (004)/(220) the noise is doubled compared with {111}, and that {111} reflections give the best result from GPA in this case. The (002)/(220) pair gives a somewhat different strain map. The (002) reflection in InGaAs/GaAs is composition sensitive, while the (004) reflection is strain sensitive. In principle, the (004)/(220) pair is a good choice since these two reflections are orthogonal to each other and thus provide the most information. However, the problem with the use of (004)/(220) in the case of Fig. 2 is noise. A single STEM image is too noisy for the low intensity reflections, such as (004). The scanning noise is a kind of high frequency noise and (004) is also affected. So, the overall noise of (004) in a single HAADF-STEM image is high.

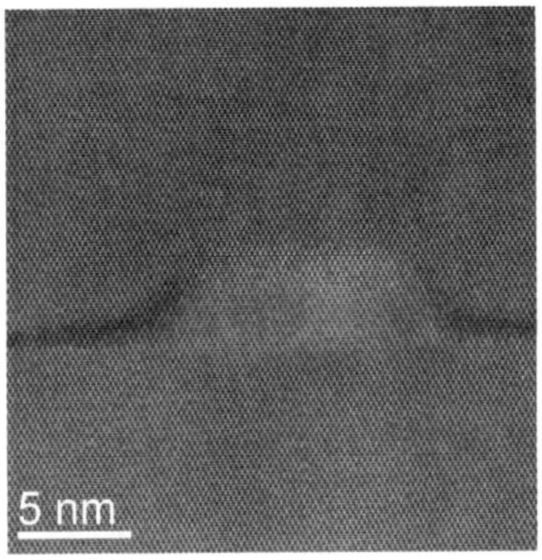

Figure 2. High resolution HAADF STEM image of InAs/GaAs QD.

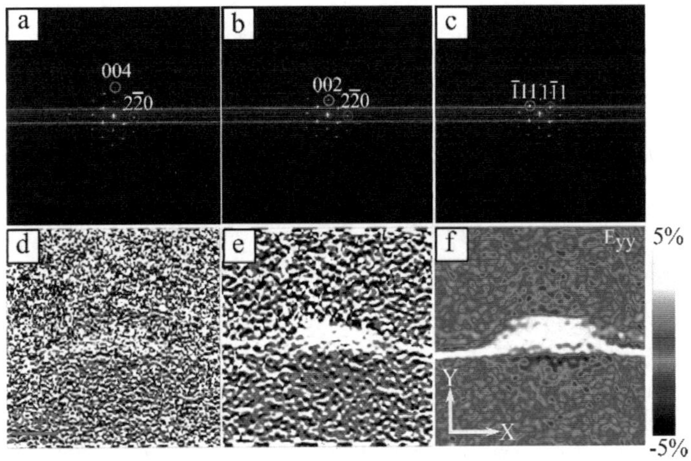

Figure 3. (d-f) ε_{yy} strain measured from Fig. 2 using the Bragg spots highlighted in the (a-c) FFTs.

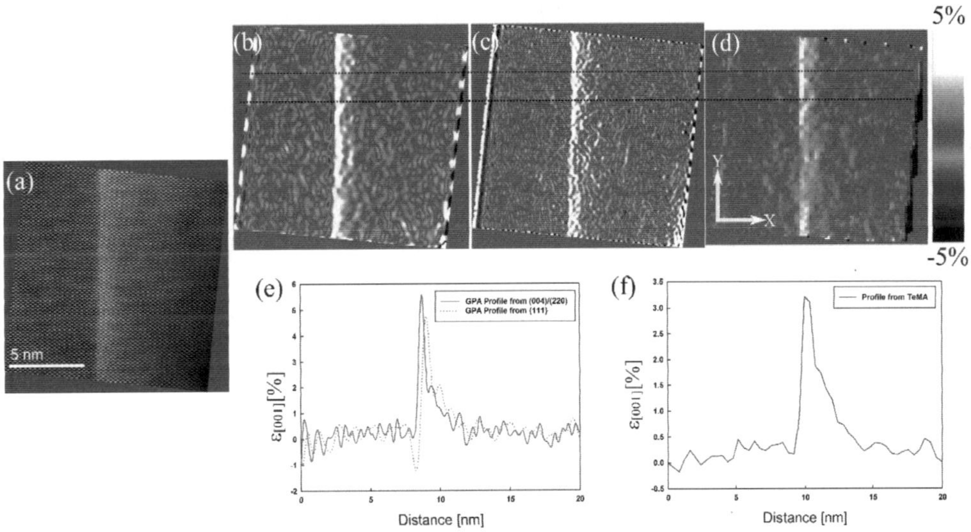

Figure 4. (a) HAADF STEM image of InAs/GaAs wetting layer after applying the SmartAlign, (b and c) ε_{xx} strain maps for strain parallel to the crystallographic [001] direction from (a) using the Bragg spots {111} and (004)/(220), respectively in the FFTs. (d) ε_{xx} strain map from TeMA, (e) The strain profile along a line crossing through the wetting layers shown with dashed lines in (b and c), (f) The strain profile along a line crossing through the wetting layers shown with dashed lines in (d).

An atomic resolution image contains multiple reflections or spatial frequencies, and GPA only uses two frequencies to calculate strain, as Fig. 3 demonstrates. While the approach used by GPA is advantageous for the analysis of HRTEM images of limited spatial resolution, ideally, the analysis of aberration corrected STEM images should take the advantages of all frequencies, and this can be achieved using real space based methods based on atomic peak finding or template matching (TeMA) [5, 6]. In these methods, the position of the atomic columns is determined by the intensity distribution, which utilizes all information recorded in the image. Fig. 4a shows a HAADF STEM image a GaAs/InAs wetting layer after applying the SmartAlign algorithm. GPA and TeMA have been used to measure the strain in this image. Fig. 4 (b and c) show the strain map from {111} and (004)/(220) reflections, respectively, and Fig. 4 (d) is the strain map from TeMA. The profiles in Fig.4 (e and f) that are taken from Fig. 4(b-d) show that (004)/(220) is in closer agreement with the real space approach (TeMA), and {111} gives systematic noise and interfacial features that are not seen in (004)/(220). This could be a Fourier termination effect due to the stronger background in the FFT spectrum near the {111} reflections.

Finally, thin foil relaxation effects are not taken into account in the strain measurements since the dimensions of the QDs are small and the sample is not extremely thin. Together, we expect the relaxation happens mostly on the same surfaces [7].

4. Conclusions

In conclusion, we have demonstrated several strategies for improving the results of strain analysis by GPA of the InAs/GaAs quantum dot (QD) material, including

1) The image quality can be significantly improved by performing multiple scans with 0° and 90° rotations and by using SmartAlign to reduce noise and scan errors.
2) The choice of *g* vectors in the GPA procedure can have a large impact on the strain mapping results. The use of {111} reflections are the most effective in case of a single HAADF-STEM image. However, the same {111} pair applied to the improved HAADF-STEM image revealed artefacts at the wetting layer interface.
3) By using SmartAlign for a series of images to improve the signals at high frequencies, and as it is shown in Fig. 4c, the (004)/(220) reflections gave the best results when using GPA for strain measurements.

The data in this paper also shows that GPA analysis using only one pair of reflections has its limitations since it does not take advantage of all information recorded at atomic resolution. For GPA, the challenge is how to combine information obtained by using different Fourier pairs in order to obtain true strain information. With aberration corrected microscopes the resolution is much higher, and by using SmartAlign we get better atomic resolution images, which definitely improves the GPA analysis. However, still additional information can be picked up with real space approaches which use information from all frequencies and would be a better approach than GPA.

Acknowledgments
The Norwegian Research Council is acknowledged for funding the HighQ-IB project under contract no. 10415201. The (S)TEM work was carried out on the NORTEM infrastructure at the TEM Gemini Centre, NTNU, Norway. JMZ is supported by Onsager Professorship of NTNU.

References

[1] Hÿtch M J, Snoeck E and Kilaas R 1998 *Ultramicroscopy* **74** 131-146
[2] Peters J J P et al. 2015 *Ultramicroscopy* **157** 91-97
[3] Shockley W and Queisser H J 1961 *Journal of Applied Physics* **32** 510
[4] Jones L and Yang H 2015 *Advanced Structural and Chemical Imaging* **1** 1-16
[5] Kim H, Meng Y, Rouviére J L and Zuo J M 2017 *Micron* **92** 6-12
[6] Nord M, Vullum P E, MacLaren I, Tybell T and Holmestad R 2017 *Advanced Structural and Chemical Imaging* **3** 9
[7] Treacy, M. M. J. and J. M. Gibson 1986 *Journal of Vacuum Science & Technology* **B 4**, 1458-1466.

IOP Conf. Series: Journal of Physics: Conf. Series **902** (2017) 012022 doi:10.1088/1742-6596/902/1/012022

Precipitate statistics in an Al-Mg-Si-Cu alloy from scanning precession electron diffraction data

J K Sunde[1], Ø Paulsen[1], S Wenner[1] and R Holmestad[1]

[1]Department of Physics, Norwegian University of Science and Technology (NTNU), Høgskoleringen 5, N-7491 Trondheim, Norway

E-mail: jonas.k.sunde@ntnu.no

Abstract. The key microstructural feature providing strength to age-hardenable Al alloys is nanoscale precipitates. Alloy development requires a reliable statistical assessment of these precipitates, in order to link the microstructure with material properties. Here, it is demonstrated that scanning precession electron diffraction combined with computational analysis enable the semi-automated extraction of precipitate statistics in an Al-Mg-Si-Cu alloy. Among the main findings is the precipitate number density, which agrees well with a conventional method based on manual counting and measurements. By virtue of its data analysis objectivity, our methodology is therefore seen as an advantageous alternative to existing routines, offering reproducibility and efficiency in alloy statistics. Additional results include improved qualitative information on phase distributions. The developed procedure is generic and applicable to any material containing nanoscale precipitates.

1. Introduction

6xxx series Al(-Mg-Si(-Cu)) alloys form an important group of engineering alloys and are found in a range of applications in buildings and constructions, transportation and marine structures. Increasing industrial demand for materials combining properties such as low weight, high strength, high formability and good corrosion resistance, make Al-based alloys prime candidates for future applications. As an example, this is reflected in their growing impact in the automotive industry [1].

The 6xxx series alloys are characterised by a significant increase in hardness upon short-term thermal ageing. This increase is caused by a large number of small, semi-coherent and metastable needle-shaped precipitates formed along the $\langle 100 \rangle_{\mathrm{Al}}$ directions in the Al matrix from a solid solution of Mg and Si [2]. These nanostructures act as obstacles to dislocation movement through the Al matrix, hence strengthening the material.

The early-stage precipitation sequence in Al-Mg-Si-Cu alloys is normally given as [3–5]

$$\mathrm{SSSS} \rightarrow \text{atomic clusters} \rightarrow \text{GP-zones} \rightarrow \beta'', \mathrm{L} \rightarrow \beta'(\text{-Cu}), \mathrm{U1}, \mathrm{U2}, \mathrm{C}, \mathrm{B}', \mathrm{Q}'.$$

The number and type of the different phases forming throughout the Al matrix are crucially dependent on alloy composition and all prior thermo-mechanical processing. In order to further optimise material properties, detailed and reliable precipitate statistics and its change versus different ageing times, is needed. This is generally acquired through bright-field (BF) and/or dark-field (DF) transmission electron microscopy (TEM) imaging techniques, with subsequent manual counting and measurements [6]. With recent developments of TEM techniques yielding large, multi-dimensional datasets [7], and corresponding parallel advancements in powerful data processing tools [8], techniques for obtaining the statistics in more objective and reproducible manners are sought. This forms the main motivation behind this work.

2. Materials and methods

The material studied has a nominal composition Al-1.11Mg-0.50Si-0.16Cu (at.%) [9]. From an extruded rod with cylindrical profile (Ø 20 mm) a sample of 10 mm height was cut and solution heat treated at 540 °C for 15 min, before water quenched to room temperature. After 4 h natural ageing it was artificially aged for 6 h at 200 °C, yielding a slightly over-aged condition. TEM samples were made by standard electro-polishing using a Struers Tenupol-5.

Scanning precession electron diffraction (SPED) was performed using a NanoMEGAS DigiSTAR scan generator fitted to a JEOL 2100F TEM operated at 200 kV, with a precession angle of 1° and a step size of 2.28 nm. To get high quality data from the nm-sized precipitates the precession system was aligned using the procedure proposed by Barnard et al. [10]. The main SPED dataset is presented as a virtual dark-field (VDF) image (Figure 1) formed by plotting the intensity of a sub-set of pixels in each PED pattern as a function of probe position. The dataset comprises 247 500 diffraction patterns (DPs) covering an area of 1.34 μm^2 in 550×450 pixels. The 4D-SPED dataset was primarily analysed using the HyperSpy [8] Python library. An image that highlights the precipitate cross-sections was obtained through principal component analysis, giving a component pattern that closely resembled the sum of precipitate-characteristic reflections. After applying local intensity thresholding, the number of precipitate cross-sections in the scanned area was obtained using a blob-detection algorithm based on the Laplacian of Gaussian method on the pre-processed image. Information on precipitate types was extracted using non-negative matrix factorisation (NMF) with a navigation mask surrounding all precipitate cross-sections detected (Figure 2). The NMF component patterns were compared to (kinematically) simulated DPs of reported phases in the Al-Mg-Si-Cu system. High-angle annular dark-field scanning TEM (HAADF-STEM) images were acquired on a double corrected JEOL ARM200CF microscope operated at 200 kV.

The total area analysed by conventional methods covered approximately 1.1 μm^2, comparable to the SPED scan area, and was acquired in the same grain. The precipitate number density, ρ [μm^{-3}], was calculated as [6]

$$\rho = \frac{3N_{\|}}{A(t + \langle l \rangle)}. \tag{1}$$

Here, $N_{\|}$ is the number of precipitate cross-sections in the imaged area, A, while t is the area thickness and $\langle l \rangle$ is the average needle length. The factor 3 accounts for the 3 equivalent $\langle 100 \rangle_{Al}$ directions in which the needles grow, hence assuming an isotropic distribution. The sample thickness, measured by electron energy loss spectroscopy (EELS), gives the highest contribution to the uncertainty in the calculated precipitate number density. The total precipitate number density, $\langle \rho \rangle$, is given as the average of all number densities calculated from single images.

3. Results and discussion

The statistics obtained using conventional and SPED methodologies are presented in Table 1, showing a number density of $(65\,700 \pm 6700)\,\mu m^{-3}$ and $(67\,800 \pm 5500)\,\mu m^{-3}$ for the two approaches, respectively. Considering the high number densities involved, this shows a close agreement between the two methods. It implies that the PED pattern stack has sufficient signal-to-noise ratio to give a number density comparable to that of broad-beam illumination modes.

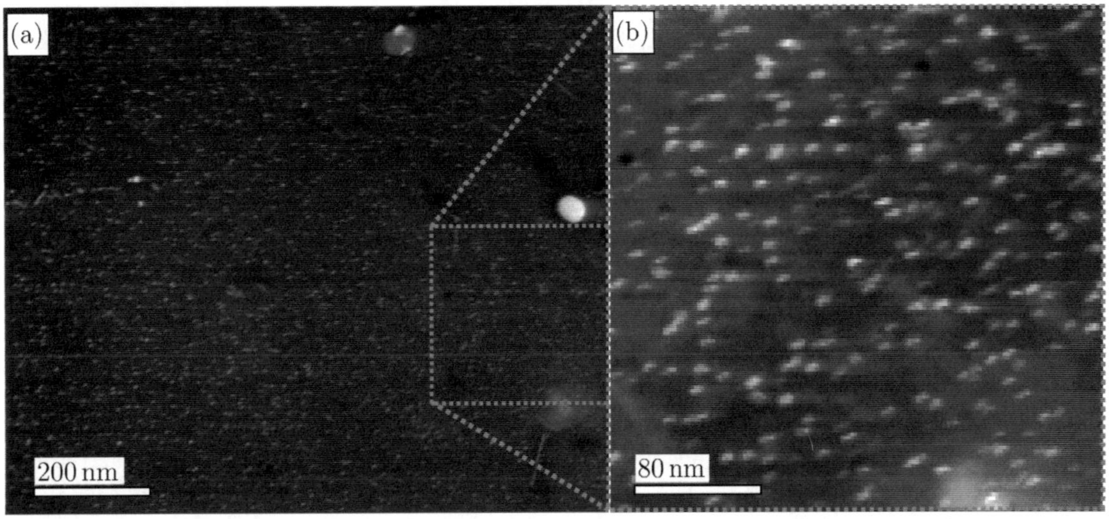

Figure 1. (a) VDF image formed from a SPED scan acquired near a $\langle 100 \rangle_{Al}$ zone axis using precipitate diffraction spots. The area covers 1.34 μm^2 in 550 × 450 pixels. (b) Enlarged view of the dashed region in (a).

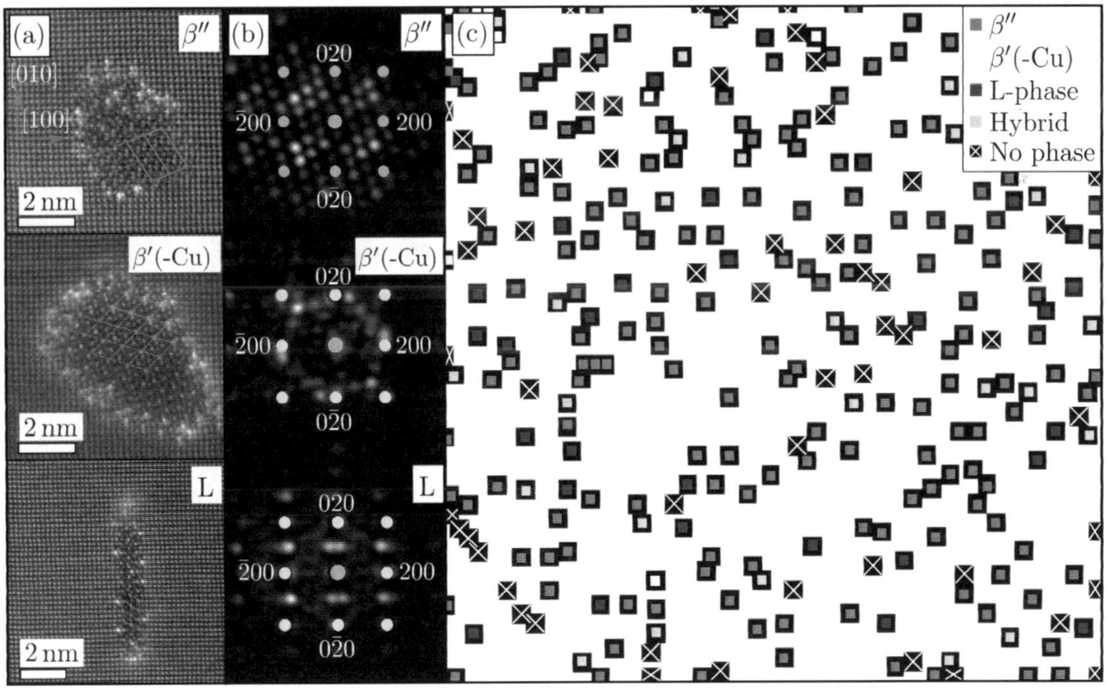

Figure 2. (a) HAADF-STEM images of the three main precipitates observed. Unit cells of the phases are highlighted in red. In the disordered L-phase, parts of the Si-network are marked. (b) NMF component patterns resembling (kinematical) DPs of the corresponding precipitate types in (a). Some aluminium reflections are marked and indexed. (c) Colour-map showing the precipitate phase distribution obtained from loading maps associated with the main NMF component patterns of (b). The area in (c) is the same as in Figure 1 (b).

The transition from Figure 1 (b) to Figure 2 (c) shows how the initial coarse estimate of number density is refined and that in addition, phase fractions are obtained. The latter has excluded those cross-sections which are not associated with a recognisable DP from the phases present. This second refinement is not attainable in manual analysis of BF/DF images where it may be difficult to differentiate between e.g. in-plane and normal orientations of small β'' needles.

Table 1. Precipitate statistics obtained using the two presented methodologies.

Method	β'' [%]	β'(-Cu) [%]	L [%]	Hybrid phase [%]	$\langle\rho\rangle$ [μm^{-3}]
Conventional	54	Unattainable	10	36	65 700±6700
SPED	73	1	12	14	67 800±5500

The precipitate fractions in Table 1 show significant variations, agreeing only on the fractions of the L-phase. This was expected, as the conventional methodology determines precipitate type solely based on shape and orientations relative to the Al matrix, while SPED uses information from reciprocal space. A phase such as β'(-Cu) does normally not extend to the Al interface, and the outer region consists of a narrow layer of other phases (or fractions thereof). Purely geometric considerations must therefore classify less well-defined cross-sections as hybrid. The L-phase has a distinct, narrow cross-section, easily recognised in BF images. Furthermore, since it rarely couples with the other phases, its fraction should be reliable.

4. Conclusions

Computational analysis of SPED data yields precipitate number densities in close agreement to conventional methodologies. Obtained phase fractions show significant deviations between the methods, and is suspected to be a consequence of inaccuracy in traditional routines being based on purely geometric considerations. With implications of equal sensitivity in the two approaches, the SPED estimate of phase fractions is considered more reliable. Due to the increased objectivity of the new analysis, the results are more readily reproducible, which has important implications for the validity and use of TEM statistics in future alloy developments. Further work aims at refining the methodology, making all steps automated and quantifying uncertainties more accurately.

Acknowledgments

All authors extend their gratitude to collaborating researchers Duncan N Johnstone and prof. Paul A Midgley from the University of Cambridge, and prof. Antonius T J van Helvoort from the Norwegian University of Science and Technology.

References

[1] Miller W S, Zhuang L, Bottema J, Wittebrood A J, De Smet P, Haszler A and Vieregge A 2000 *Mater. Sci. Eng.* A **280** 37–49
[2] Totten G E and MacKenzie D S 2003 *Handbook of Aluminium* (New York: Marcel Dekker)
[3] Edwards G A, Stiller K, Dunlop G L and Couper M J 1998 *Acta Mater.* **46** 3893–904
[4] Marioara C D, Andersen S J, Stene T N, Hasting H, Walmsley J, van Helvoort A T J, and Holmestad R 2007 *Phil. Mag.* **87** 3385–413
[5] Ravi C and Wolverton C 2004 *Acta Mater.* **52** 4213–27
[6] Andersen S J 1994 *Metall. Mater. Trans.* A **26** 1931–37
[7] Rauch E F, Portillo J, Nicolopoulos S, Bultreys D, Rouvimov S and Moeck P 2010 *Z. Kristallogr.* **225** 103–9
[8] de la Peña F et al. HyperSpy-1.3 2017 DOI: https://doi.org/10.5281/zenodo.583693
[9] Marioara C D, Andersen S J and Røyset J 2014 *Metall. Mater. Trans.* A **45**, 2938–49
[10] Barnard J S, Johnstone D N and Midgley P A 2017 *Ultramicroscopy* **174** 79–88

Formation of Nanosized Lamellas of a Hardening Intermetallic Phase in the Powder Ni-based Coating Deposited by Microplasma Spraying on Steel Substrates

D Alontseva[1], E Ghassemieh[2], A Russakova[1], A Dzhes[1] and N Prokhorenkova[1]

[1]Department of Instrument Engineering and Technology Process Automation, East-Kazakhstan State Technical University, 69 Protozanov St, Ust-Kamenogorsk, 070004, Kazakhstan
[2]Faculty of Engineering and Advanced Manufacturing University of Sunderland, St. Peters Campus, Sunderland, SR6 0AN

E-mail: dalontseva@mail.ru

Abstract. This paper presents new results of studying the structure-phase compositions of Ni-based powder coating, deposited by Microplasma Spraying onto steel substrate by transmission electron microscopy (TEM), scanning electron microscopy (SEM), and X-ray diffraction (XRD). It is shown that we have managed to obtain the predicted specific structure with nanosized lamellas of intermetallic phases due to the appropriate proper selection of modes of additional microplasma processing. As a result, the microhardness of the coating has been increased by 1.25 times.

1.Introduction

Among the various existing plasma spraying processes for surface protection against corrosion and wear, Microplasma Spraying (MPS) is very beneficial for the deposition of coatings on small parts or with high accuracy [1,2]. However, a number of challenges still remain in the field of microplasma coating. The most important among them is the problem of the formation of coatings with specified structure and properties. Usually we have to use combined methods of plasma spraying including the deposition of powder on the substrate and the additional treatment of the coated surface by irradiation. Additional processing of the powder plasma coatings by irradiation is used to melt the coating in order to reduce its roughness and increase the homogeneity of its structure. This is reported in previous work [3,4]. We also proposed an alternative approach to the choice of additional processing as shown in our past works [5,6]. The selection of additional treatment modes has been based on the results of simulation of temperature fields in the "coating-substrate" system under heating by a moving plasma source. We have indicated the precipitation in the coating during additional irradiation of the phases with defined morphology and composition that could increase the coating's microhardness.

The purpose of this study is to use microplasma and additional surface treatment to form a nickel based coating with the desired microstructure, i.e. lamellas of the intermetallic phase.

2. Materials and Methods

The 100 µm thick coatings from PG-19N-01 powder (Ni-based powder alloy with addition of Cr (14…20%), B (3.5%), Si (4.3%), Fe (7%), C (0.8%)) were deposited on Steel 3 substrates (Fe – base, C - 0.25 %, Mn – 0.8 %, Si – 0.37 %, P < 0.045%) by MPS.

The microplasma deposition of powders onto steel substrates was carried out with the help of "MPN-004" micro-plasma deposition unit (produced by E. O. Paton Electric Welding Institute, Ukraine). The powder is fed in a stream of argon onto a substrate.

The modes of microplasma deposition: a laminar plasma jet; the diameter of the spray spot (1…8) mm; the power of plasma source 2 W; the powder flow rate 2 kg/hour; the travel speed of the plasma jet 0.008 m/s

The additional treatment of the samples by a plasma jet was carried out at power density of $2.0 \cdot 10^9$ W/m^2 with 0.006 m/s of the plasma jet travel speed.

Experimental methods of analysis include Transmission Electron Microscopy (TEM) by JEOL JEM-2100 with Energy Dispersive Spectrometry (EDS) INCA Energy TEM 350, Scanning Election Microscopy (SEM) by JEOL JSM-6390LV, X-ray diffraction (XRD) by X'Pert PRO. Gatan M-691 Precision Ion Polishing System was used to prepare TEM foils by the Ar ion sputter etching method. Microhardness test of the samples was performed with LM-700 digital microhardness meter.

3. Results and Discussion

The XRD results shows up to about 10 vol.% of $CrNi_3$ phase with fcc crystal lattice (a = 3.55 Å) is formed. This is whilst the original powder coating does not contain this phase (figure 1).

The microhardness of the coating in average is 6.0± 0.5 GPa, and that of the substrate is 1.4 ± 0.1 GPa. After additional pass by a plasma jet the $CrNi_3$ phase volume fraction increased by 5 vol. % of (figure 1), and the coating microhardness increased by 7.5 GPa (figure 2).

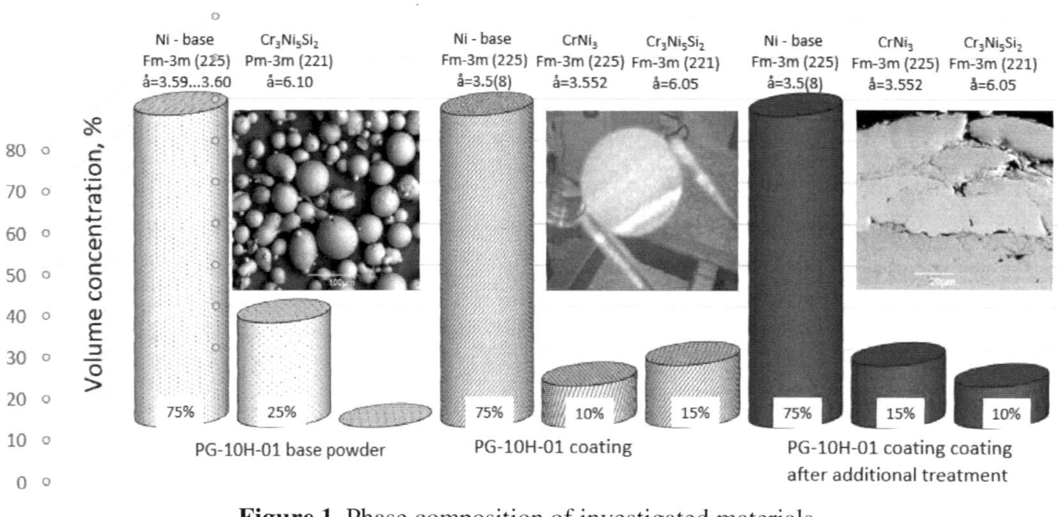

Figure 1. Phase composition of investigated materials

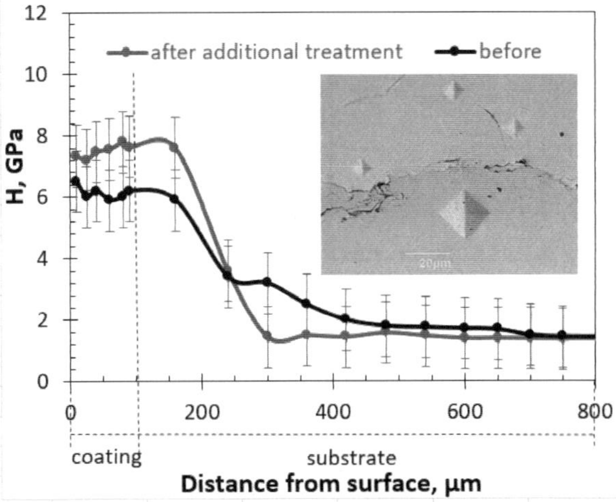

Figure 2. Coatings microhardness variation according to distance from the surface

The analysis of the coating by TEM method showed that the $CrNi_3$ phase is released in the form of nanoscale lamellae (figure 3a) from the fcc coating matrix (γ-phase, solid solution on Ni-base with parameter a = 3.5(8) Å) consisting of crystallographically disoriented, approximately equiaxed grains with a diameter of about 50 nm (figure 3b).

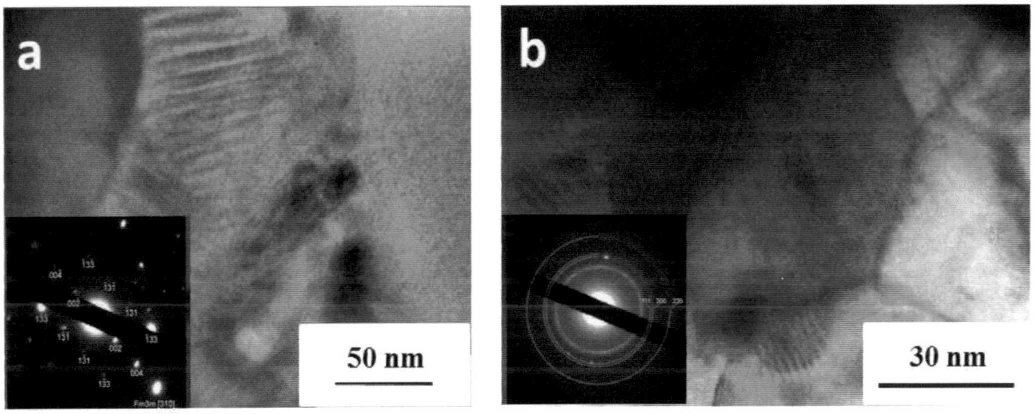

Figure 3. TEM images (with corresponding diffraction patterns) of lamellas of **$CrNi_3$** -phase (a); of nanograins with different crystal-lattice orientation (b) in the Ni-based coating

These results are of significance due to the followings: firstly, the coating consists of the ductile base with fcc lattice type which is reinforced by lamellas of hard and heat-resistant intermetallic phase. It gives the possibility of creating a protective coating having a combination of properties such as thermal stability, high hardness and good ductility at the same time. This is also noted in the work of other researchers [1,2]. Secondly, we proposed an alternative approach to the choice of modes of additional processing. As the authors of papers [3, 4] have noted, an additional treatment of powder plasma coatings by irradiation results in melting of the coating. This consequently improves the homogeneity and adhesion to the substrate results in the melting of the coatings. The use of relatively low power densities of the plasma source in our study allowed the hardening of the coating due to the phase transformations which take place at lower temperatures. However, our proposed approach is not universal and requires a thorough and further analysis of the coating material and desirable and possible phase transformations.

4. Conclusions

The laboratory samples with protective powder coatings deposited by the microplasma according to the recommended modes onto steel substrates have been obtained; and it was established experimentally that the coatings have the predicted structure-phase composition, namely the nanograin Ni-based solid solution with precipitations of strengthening intermetallic lamellas of $CrNi_3$ -phase which provide the coating's high microhardness.

TEM and XRD methods confirmed a 4% increase in the volume fraction of $CrNi_3$ phase in the coating after additional microplasma irradiation.

It is established that microhardness of the coatings after additional processing is 1.25 times higher than that of the same coatings before irradiation, and 5 times higher than that of the substrate.

References

[1] Lugscheider E, Bobzin K, Zhao L, Zwick J 2006 *Assessment of the Microplasma Spraying Process for Coating Application* Advanced Engineering Materials Special Issue: Thick Coatings for Thermal, Environmental and Wear Protection Volume 8 Issue 7 p 635–639

[2] Travis L, Michael J 2013 *J. Vac. Sci. Technol.* A **31** 061312

[3] Vardelle A, Moreau Ch, Nickolas J, Themelis A 2015 *Plasma Process* **35** 491–509

[4] Serres N, Hlawka F, Costil S, Cornet A, Langlade C. et. al. 2010 *Journal of Optoelectronics and Advanced Materials* **12** (3) 505-510

[5] Alontseva D., Krasavin A, Russakova A., Kolesnikova T, Bektasova G 2016 *Medžiagotyra* **22** (2) 38-242

[6] Alontseva D, Ghassemieh E 2015 The Structure-Phase Compositions of Powder Ni – based Coatings after Modification by DC Plasma Jet Irradiation IOP Publishing: Journal of Physics: Conference Series 644 012009

Acknowledgments: The authors thankfully acknowledge the funding provided by the Scientific Committee of the Ministry of Education and Science of the Republic of Kazakhstan for the project 0869/GF titled "Automated precision deposition of powder coating multifunctional systems"

A Hydrogenated Amorphous Silicon (a-Si:H) Thin Films for Heterojunction Solar Cells: Structural and Optical Properties

Ayse Seyhan[1,2], Tolga Altan[2], Ömer Can Ecer[1,2], Recep Zan[2]

[1]Department of Physics, Nigde Omer Halisdemir University, Nigde-Turkey
[2] Nanotechnology Application and Research Center, Nigde Omer Halisdemir University, Nigde-Turkey
Email: aseyhan@ohu.edu.tr

Abstract: Heterojunction solar cells composed of hydrogenated amorphous silicon (a-Si:H) and c-Si has been widely studied due to its excellent photovoltaic characteristics. In this study, we studied the structural and optical properties of a-Si:H thin films for heterojunction solar cells by using SEM, TEM and ellipsometry. We found that uniform deposition of a-Si:H thin film on substrate is crucial to reach high solar cell efficiencies.

1. Introduction

Current photovoltaic trends shows that both improved efficiency and lowered cost can be obtained by heterojunction solar cell technology [1]. Silicon heterojunction solar cells consist of crystalline silicon (c-Si) wafers with typically 3-15 nm thick amorphous silicon layers deposited on both sides to form electrically passivated contacts. A hydrogenated i/p layer stack forms the emitter at the front side and a hydrogenated i/n layer stack forms the back-surface field. The use of such layers in heterojunction cells has many advantages such as low process temperatures, low cost, large area deposition and high cell efficiencies. Such a device structure has be pioneered by Sanyo [2] starting in the early 1990s increasing efficiency steadily to currently 24.7% [3] and 26.6% [4] with a double-sided heterojunction and an IBC-SHJ cell, respectively.

Here we present a scanning electron microscopy study of the a- Si:H/c-Si- Heterojunction solar cell which has revealed that the uniform deposition of thin films has critical importance to get high cell efficiency.

2. Experimental

In order to obtain a-Si:H thin films, the plasma enhanced chemical vapor deposition (PECVD) technique has been used, which is well known for thin film growth [5]. The a-Si:H layers were deposited in a radio frequency (13.56 MHz) PECVD reactor. The temperature for the cell fabricating processes was around 200°C. The a-Si:H precursor gases were semiconductor grade silane (SiH_4), phosphine (PH_3) diluted to 1 % in hydrogen, trimethlyboron (TMB) diluted to 1% in H_2 and 5N H_2.

The structure of the 156 mm x156 mm a- Si:H/c-Si -Heterojunction solar cell is shown in Figure 1. First n-type c-Si wafers were cleaned with HF just before the deposition. Intrinsic a-Si:H layer (5-10nm) deposited in one reactor and then n (p) type a-Si:H layer (5-10 nm) deposited in an another reactor to get rid of contamination. Both ITO and back contact Ag were deposited by Physical Vapor Depositon (PVD). A front grid was printed with silver (Ag) paste. All these steps were characterized by spectroscopic ellipsometer and profilometer. All layer properties of Si heterojunction solar cell have been studied by SEM (HITACHI SU5000, for front side of solar cell 100 k magnification, obtained at 5.0 kV; for back side of solar cell 11 k magnification, obtained at 10.0 kV).

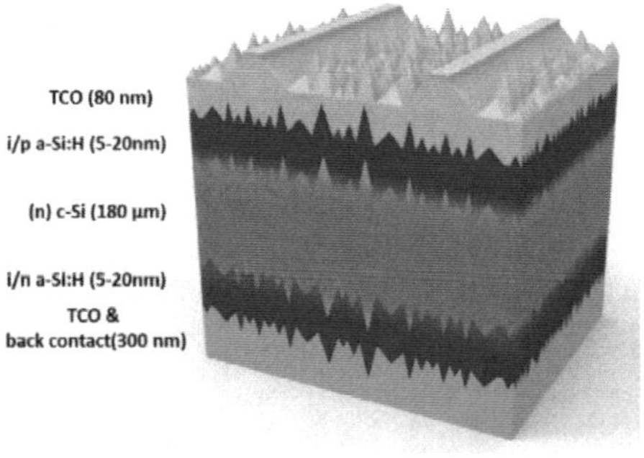

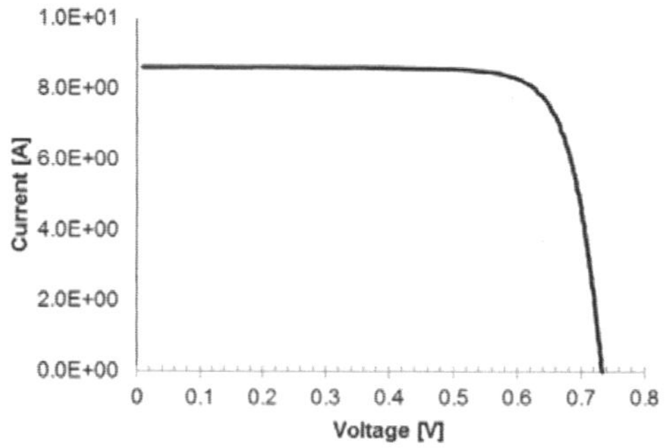

Figure 1. The structure of the a-Si:H/c-Si heterojunction solar used in this (left). Current-voltage (I-V) performance of optimized a- Si:H/c-Si heterojunction solar cell obtained on commercial 156 mm pseudo-square n-type Cz c-Si substrate (right).

3. Result and Discussion

It is well-known that by adding the high-quality i-a-Si:H layer, the defect density of the a-Si:H/c-Si interface can be reduced, resulting in a high conversion efficiency, high V_{oc} and excellent temperature coefficient [6,7]. However, the surface morphology on the textured substrate must be considered in order to optain high conversion efficiency. There can be no doubt that the quality and uniformity of thin films on texturized wafer dominates the performance of the a-Si:H/c-Si heterojunction solar cell.

a-Si:H thin films have been deposited on textured c-Si wafer with different thickness which are 5, 10, 15, and 20 nm. The deposition of high quality a-Si:H layers reduced surface recombination, thus improving the junction properties. V_{oc} values were improved with increasing a-Si:H layer thickness. However, samples with thicker layer shown a large decrease in FF which can be attributed to an increase in series resistance. The a-Si:H thin films with different thickness were deposited quite uniformly on the c-Si wafer. The results summarized table 1.

Table 1. Solar cell characretistics as a function of a-Si:H thickness. Each data represents the average value of some cells.

PV parameters	**a-Si:H layer thickness**			
	5 nm	**10 nm**	**15 nm**	**20 nm**
V_{oc}(mV)	680.0	700.0	715.7	715.0
J_{sc} (mA/cm^2)	35.7	36.2	36.3	36.2
FF (%)	78.5	78.0	77.0	76.8
Efficiency(%)	19.2	19.7	19.9	19.6

The SEM images and EDS mapping of the front and back side contacts are shown in Figure 2(a-b) and Figure 3, respectively. The front contact was fabricated by PVD- 80nm thick ITO film- on the top side of the a-Si:H/c-Si heterojunction solar cell. The rear contact was then deposited (ITO-50nm/Ag 240nm) on the back side of the solar cell. The thin films composition can be seen clearly on the EDX mapping.

It can be clearly observed in the SEM images (figure 2) that the thin film thickness is quite uniform all over the textured surface of a-Si:H/c-Si heterojunction solar cells. EDS maps in figure 3 also provide a clear distinction between deposited thin films.

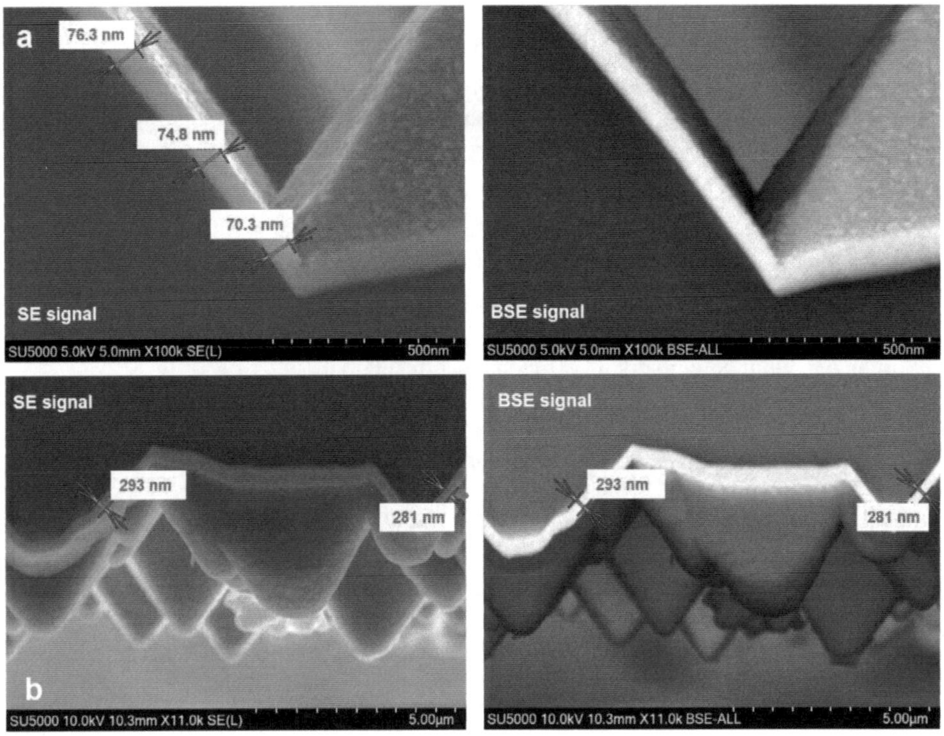

Figure 2. (a) Front side, and (b) Back side SE and BSE HR-SEM image of a- Si:H/c-Si -Heterojunction solar cell at 100 k magnification, obtained at 5.0 kV, at 11 k magnification, obtained at 10.0 kV, respectively.

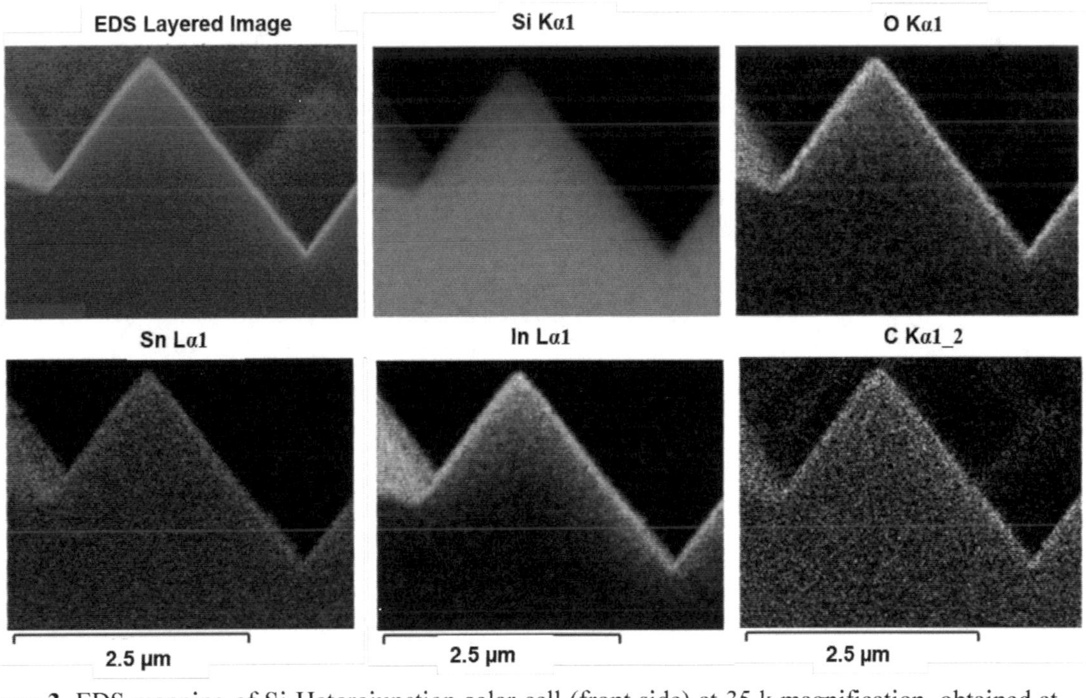

Figure 3. EDS mapping of Si-Heterojunction solar cell (front side) at 35 k magnification, obtained at 10.0 kV.

The photovoltaic properties of the a- Si:H/c-Si heterojunction solar cell were studied by using current-voltage (I-V) measurements. The heterojunction solar cell showed a good photovoltaic behaviour with energy conversion efficiency 20% on 6" n-type c-Si. The obtained photovoltaic parameters are found to be open circuit voltage Voc = 715.7 mV, short circuit density Jsc = 36.3 mA/cm^2, and fill factor FF = 77%, Figure 1 (right).

4. Conclusion

In conclusion, we used an SEM operated at 5 and 10 kV to investigate thin films uniformity in Si a-Si:H/c-Si heterojunction solar cells. The SEM images reveal that all thin films on the heterojunction solar cell are deposited uniformly. This shows that the optimized thin film growth parameters have a critical importance to get high conversion efficiency from solar cells.

References

[1] Antoine Descoeudres, Zachary C. Holman, Loris Barraud, Sophie Morel, Stefaan De Wolf, and Christophe Ballif, IEEE Journal of Photovoltaics, 2013, Vol. 3, No. 1.
[2] M. Taguchi,M. Tanaka, T. Matsuyama, T. Matsuoka, S. Tsuda, S. Nakano, Y. Kishi, and Y. Kuwano, Proc. Int. Photovoltaic Sci. Eng. Conf.-5 Tech. Dig., 1990, pp. 689–692.
[3] Mikio Taguchi, Ayumu Yano, Satoshi Tohoda, Kenta Matsuyama, Yuya Nakamura, Takeshi Nishiwaki, Kazunori Fujita, and Eiji Maruyama, IEEE Journal of Photovoltaics, Vol. 4, No. 1, January 2014.
[4] Kaneka Corporation, 2016.
http://www.kaneka.co.jp/kaneka-e/images/topics/1473811995/1473811995_101.pdf
[5] L. Korte, E. Conrad, H. Angermann, R. Stangl, M. Schmidt, Solar Energy Materials & Solar Cells 2009, 93, 905–910.
[6] Seif J P, Krishnamani G, Demaurex B, Ballif C, Wolf S D. Amorphous/crystalline silicon interface passivation: ambient-temperature dependence and implications for solar cell performance. IEEE Journal of Photovoltaics, 2015, 5(3): 718–724,
[7] Geissbühler J, De Wolf S, Demaurex B, Seif J P, Alexander D T L, Barraud L, Ballif C. Amorphous/crystalline silicon interface defects induced by hydrogen plasma treatments. Applied Physics Letters, 2013, 102(23): 231604.

Acknowledgements
We would like to acknowledge funding from Turkish Ministry of Development and we would like to thank for the data to HITACHI.

Fabrication and Characterisation of an Adaptable Plasmonic Nanorod Array for Solar Energy Conversion

J Cottom[1], P Abellan[2], FS Hage[2], QM Ramasse[2], K Critchley[3], R Brydson[1]

[1] School of Chemical and Process Engineering, University of Leeds, UK
[2] SuperSTEM Laboratory, Daresbury, UK
[3] School of Physics and Astronomy, University of Leeds, UK

E-mail: pm08jc@leeds.ac.uk

Abstract. The surface plasmonic modes of a side-by-side aligned gold nanorod array supported on a gold substrate has been characterised by electron energy loss spectroscopy (EELS). Plasmonic coupling within the array splits the nanorods' longitudinal mode into a bright mode (symmetrically aligned dipoles) and a dark mode (anti-symmetrically aligned dipoles). We support this observation by means of finite element modelling (FEM).

1. Introduction

Metallic nanoparticles exhibit fascinating optical properties based on surface plasmon resonances (SPRs) which could aid in the design of advanced metamaterials. The field of plasmonics has received much attention due to the generation of localized electromagnetic fields with enhanced strengths and increased light absorption. With careful design of the geometry, the material and the distance between nanoparticles, a device may be engineered to enable a variety of applications including molecular detection using surface enhanced Raman spectroscopy (SERS) [1] and photocatalysis [2].

Coupling between SPRs of neighbouring nanoparticles is of interest as it can provide large field enhancements localised between the particles and dramatic shifts in resonance energy. Such a coupling-induced spectral shift has been explained theoretically within the plasmon hydridisation model [3]. This is analogous to molecular orbital theory and, for a strongly coupled nanorod dimer, this predicts two possible arrangements of the dipole moments which give rise to two coupled SP dipolar modes. For a symmetric alignment, the modes have a net dipole moment and may be excited by optical means, referred to as "bright modes". For an anti-symmetric alignment, there is a zero net dipole moment, making these modes in effect "invisible" to light and hence denoted as "dark modes". In contrast, electron spectroscopy techniques (such as EELS) are capable of probing both bright and dark modes of a nanorod array and the spatial and energy resolution achievable in modern monochromated aberration-corrected analytical electron microscopes allow the spatial distribution of these modes to be visualized. The aim of this work is both to understand the fundamentals involved in plasmonic coupling between nanorods and to investigate the design of high surface area devices based on nanorod arrays coated with thin semiconductor layers, for efficient solar energy conversion.

2. Methodology

Gold nanorod arrays (figure 1) were fabricated by electrodeposition within porous anodic aluminium oxide (AAO) templates [4]. The AAO templates were initially produced by a 2-step anodisation method whereby 0.5 mm thick aluminium (99.9999% purity) supplied by Alfa Aesar was anodised in a 0.3M H_2SO_4 electrolyte at a voltage of 25V. This was anodised for 24 hours producing nanoscale pores that gradually ordered into a regular hexagonal arrangement with time. This first alumina layer was then selectively etched using a solution of 6wt% phosphoric acid and 1.5wt% chromium (VI) oxide at a temperature of 65°C for two hours. Although this removed the alumina layer, the remaining aluminium was textured with a highly ordered pattern from the previously grown pores. Performing anodisation a second time selectively starts the pore growth at these sites leading to an ordered array of pores that propagate vertically down through the alumina. Unreacted aluminium was then selectively etched away using a solution of 0.1 M copper (II) chloride and 6M HCl, before then also removing the insulating barrier layer by soaking the AAO in 5wt% phosphoric acid for 20 minutes. This resulted in an AAO template with ordered pores of approximately 30nm in diameter.

The nanorod arrays were grown in these pores by first thermally evaporating a 150 nm thick gold layer on the bottom of the AAO to form a working electrode. This was then placed in a three-electrode cell using a platinum wire as the counter electrode and an Ag/AgCl reference electrode. A gold electrolyte supplied by Technic Inc. (Orotemp 24 RTU) was then added and subsequently electrodeposited within the pores using an applied potential of -0.92 V versus the reference electrode. The charge passed was monitored to control the aspect ratio (AR) of the nanorods.

An electron transparent (<100 nm thick) cross-section sample of the embedded nanorod array and substrate was prepared for STEM-EELS investigations by ion beam thinning in a Gatan PIPS; the AAO template was then partially removed by etching in a 0.5M NaOH solution for 4 minutes. All EELS measurements were performed at the SuperSTEM facility using a Nion UltraSTEM 100MC "Hermes" with Nion Quadrupole-Octupole (QO) aberration corrector and a monochromated electron source. This allowed for a simultaneous sub-Angstrom electron probe size and a high energy resolution on the order of 15 meV. All EEL spectra were acquired at an accelerating voltage of 60 kV, collection and convergence angles of 44 and 31 mrads respectively and an energy dispersion of 0.01 eV per channel.

Three-dimensional finite element modelling (FEM) of optical properties was performed using Comsol Multiphysics v5.2a. A two-step approach was used whereby the solution for a p-polarised plane wave incident on a 150nm thick gold substrate was first determined in the absence of a nanorod according to Fresnel's equations. Periodic Floquet boundary conditions were applied on each side wall in order to represent the substrate as infinite. This solution was then input as the background solution with the nanoparticle present. To simulate an infinite array of nanorods, the periodic conditions were also applied in the second solution. The refractive index of gold was taken from reference [5].

3. Results

It is well known that an isolated gold nanorod exhibits both a transverse surface plasmon mode with an oscillation over the short axis of the rod and a longitudinal mode with an oscillation over the long axis. FEM of a 30 nm diameter nanorod with aspect ratio (AR) 3 in a vacuum (not shown) predicts that dipolar plasmonic modes occur in the visible region at ca. 500 nm and 600 nm for transverse and longitudinal modes respectively; the exact position of the latter depending on rod-end geometry. FEM modelling also predicts that the position of the transverse mode is relatively insensitive to nanorod diameter, whereas the longitudinal mode is weakly red-shifted with increasing rod diameter and more strongly red-shifted with increasing rod aspect ratio, with the additional appearance of higher order (e.g. quadrupolar) modes for high AR rods at shorter wavelengths. When stood vertically on a gold substrate, coupling between the nanorod and its substrate image in the metallic film induces a strong red-shift of the longitudinal mode, analogous to coupling seen in an end-to-end configuration of nanorods [3].

In addition to the particle-substrate coupling, nanorod dimers have also been shown to couple if the separation is sufficiently small [6]. Based upon plasmon hybridisation theory, this coupling is predicted to result in the splitting of the longitudinal mode into both a bright and dark coupled mode. For side-by-side aligned nanorods, the dark mode is calculated to occur at lower energy and has an anti-symmetric alignment of the dipole moments leading to an attraction of the surface plasmons. On the other hand the bright mode occurs at higher energy and has a symmetrical alignment of the dipole moments and therefore the surface plasmons are expected to repel.

Experimental EEL spectra and corresponding spectrum image (SI) maps for a gold nanorod dimer stood vertically on a 150 nm thick gold substrate are shown in Figure 2. Although only two nanorods are shown, these formed part of a much longer nanorod array. The nanorods have a diameter and aspect ratio (AR) of approximately 30 nm and AR=3 respectively and are separated by approximately one nanorod diameter. As stated above, single unsupported nanorods of an equivalent size would exhibit a longitudinal dipolar plasmonic mode in the visible region (~600 nm). However, EELS measurements show a red-shifted mode occurring at a much higher wavelength of 1348 nm (0.92 eV) owing to both coupling with the substrate and between the nanorods. Analysis of the SI map for this energy range suggest that this mode is indeed related to the dipole mode due to the energy loss being localized towards the end of the nanorod (Figure 2b). Furthermore, the SI map of the nanorod dimer shows that the energy loss is localized towards the dimer gap suggesting that coupling between the nanorods is present and is related to the attractive dark mode. However, according to the plasmon hybridisation theory, splitting of the longitudinal mode should result in an optically bright mode in addition to the dark mode discussed above. Based upon the rod centre-to-centre separation distance, this splitting of the mode would be fairly symmetrical due to dipolar interactions being dominant [7]. This would mean a corresponding blue-shift of the bright mode of a similar magnitude to that seen for the red-shift of the dark mode (with respect to the non-hybridised longitudinal mode). The EEL spectrum in Figure 2a exhibits three peaks at higher energies than the dark mode. The mode at 2.33 eV (~530 nm) is the transverse mode, whereas the peak at 1.41 eV (880 nm) appears to be related to a defect in the right-hand nanorod, where a sharp gap is formed. The mode at 1.96 eV (~630 nm) however is distributed towards the base of the nanorods, as shown in Figure 2c. Although the 1.96 eV (~630 nm) mode has a similar field distribution to that of a quadrupolar longitudinal mode, FEM simulations detailed below suggest that this relates to the bright mode of the supported array.

FEM was performed for both individual and infinite arrays of gold nanorods supported on an Au substrate and excited by a p-polarised plane wave (exciting longitudinal bright modes). Modelling individual nanorods in the array when supported on a substrate <u>but without any inter-rod coupling</u>, the resonant energy of the longitudinal mode is predicted to occur at ca. 1000nm. As the dark mode seen with EELS actually occurs at 1348nm, symmetrical splitting of the two coupled modes suggests that the bright mode would then be expected to occur at ca. 650nm, in agreement with experimental observations. Additionally, modelling the nanorods within an infinite array shows a redistribution of the electric field from the ends of the nanorod to a localisation of the bright mode towards the base of the nanorod (Figure 3a). This occurs as the inter-particle spacing is reduced (coupling strength increased) and is concurrent with a blue-shift and reduction in intensity of the plasmon mode (Figure 3b). These results support the conclusion that both the bright and dark modes arising from coupling within a supported nanorod array have been characterized and visualized at high spatial resolution by EELS, providing insight that can aid in the development of efficient solar device applications. Ongoing FEM modelling includes the incorporation of excitation by an electron probe rather than a plane wave, allowing simulation of both bright and dark modes.

Acknowledgements
SuperSTEM is the EPSRC National Facility for Advanced Electron Microscopy. JC is funded by the EPSRC DTC in Low Carbon Technologies.

References

[1] Schlücker S. 2014 *Angew. Chem. Int. Ed* **53** 4756
[2] Liu Z et al. 2011 *Nano Lett.* **11** 1111
[3] Jain P. K et al. 2010 *Chem. Phys. Lett* **487** 153
[4] Martin C. 1996 *Chem. Mater.* **8** 1739
[5] Johnson P. B. and Christ R.W.y 1972 *Phys. Rev. B.* **6** 4370
[6] Funston A. M. et al, 2009 *Nano. Lett.* **9** 1651
[7] Willingham B. 2008 *Appl. Phys. B* **93** 209

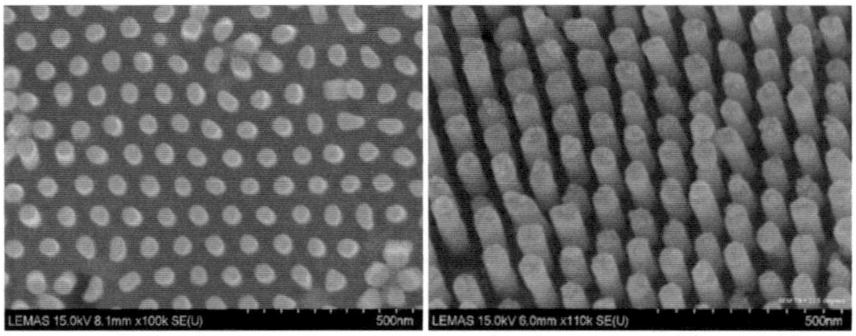

Figure 1. SEM images of a gold nanorod array grown via electrodeposition within AAO templates for a top down view (left) and tilted view (right).

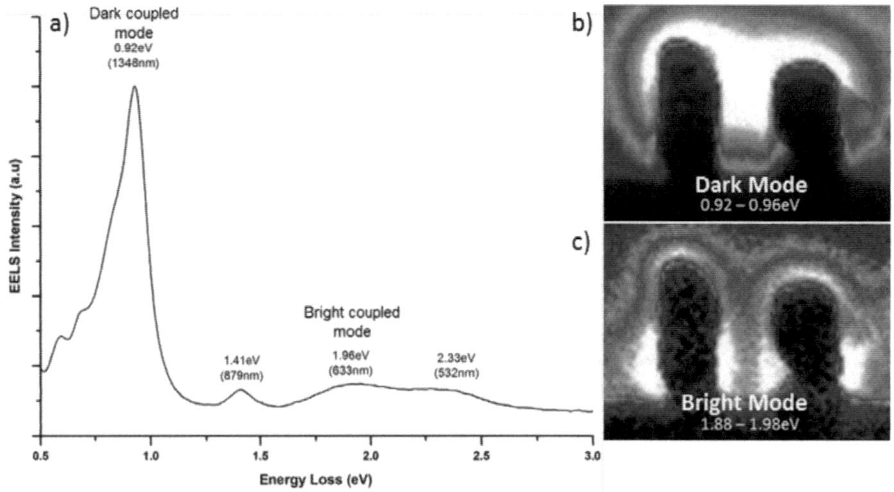

Figure 2. a) EELS spectrum for an Au nanorod dimer obtained from integrating over all the pixels in the corresponding SI. b) SI map of the dark and c) bright mode over stated energy windows.

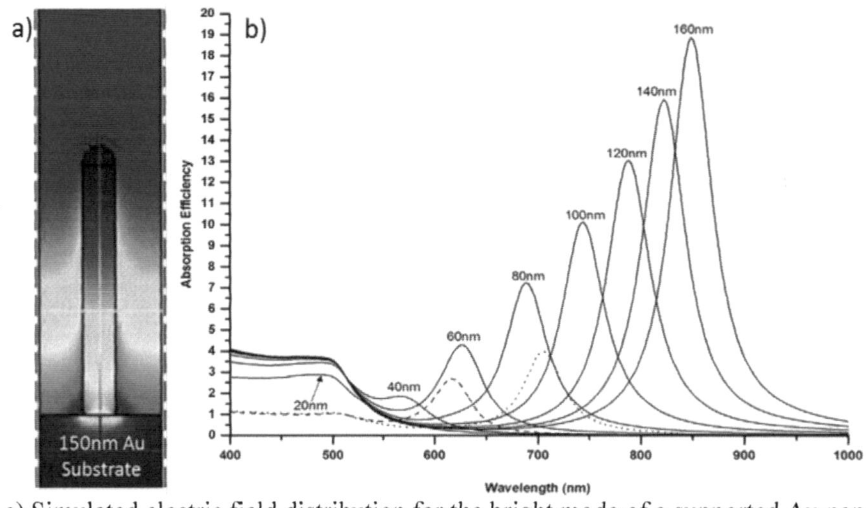

Figure 3. a) Simulated electric field distribution for the bright mode of a supported Au nanorod array (AR=7). Red-dashed lines indicate periodic boundary conditions. b) Blue-shift and reduced absorption efficiency of the bright mode with decreasing inter-rod spacing for an AR=3 supported nanorod array; dashed (AR=3) & dotted (AR=4) lines show results for single unsupported nanorods.

Determination of the lateral size and thickness of solution-processed graphene flakes

Li-Shang Lin[1], Wei Bin-Tay[1], Zabeada Aslam [1], A.V.K. Westwood[1] and R. Brydson[1]

[1] School of Chemical & Process Engineering, University of Leeds, Leeds LS29JT, UK
E-mail: pmlsl@leeds.ac.uk

Abstract. We present a method to determine the lateral size distribution of solution–processed graphene via direct image analysis techniques. Initially transmission electron microscopy (TEM) and optical microscopy (OM) were correlated and used to provide a reliable benchmark. A rapid, automated OM method was then developed to obtain the distribution from thousands of flakes, avoiding statistical uncertainties and providing high accuracy. Dynamic light scattering (DLS) was further employed to develop an in-situ method to derive the number particle size distribution (PSD) for a dispersion, with a deviation lower than *22%* in the sub-micron regime. Methods for determining flake thickness are also discussed.

1. Introduction

Graphene has gained enormous interest since its unique properties and practical isolation were demonstrated under ambient conditions [1]. While potential applications are still being discussed, developing a practicable scalable production process is a key obstacle. Solution-processed exfoliation methods are one of the most promising approaches to achieve large-scale graphene production. However these methods do not yet produce completely exfoliated graphene [2]. To monitor and optimize the graphene production process, a fast, standardized and reliable characterization protocol for large-scale solution-processed graphene is therefore desirable for both industry and academia.

The lateral size of graphene flakes is one of the most important factors affecting properties as variation in size and geometry causes a change in ratio between edge and bulk structures, resulting in spatial confinement in specific dimensions that alters the electrical and mechanical behaviour [3]. Even though characterization techniques have constantly improved, methods to obtain the lateral size distribution of solution-processed graphene are still limited owing to the difficulties in visualizing the ultra-thin nano-flakes and the fact that many of the properties of graphene are, de facto, still unknown.

Here, we demonstrate methodologies to determine the lateral size distribution for solution-processed graphene. The lateral dimension distribution was measured via direct imaging using two microscopy techniques (TEM and OM) and via a fast, but less direct technique based on DLS. Approximations, errors and deviations are calculated and discussed.

2. Experimental

The graphene sample used was synthesized by milling graphite powder in ionic liquids (2Dtech™ Aquagraph series). The graphene sample was dispersed in isopropyl alcohol (IPA) at a concentration estimated to be $1.65 \pm 0.21 (g/ml)$. A 300-second sonication (40 kHz, 80W) process was applied prior to each experiment to overcome any serious agglomeration. The resulting suspension was drop cast onto holey carbon grids and SiO_2/Si substrates (284.1 ± 0.75 (nm) thick) for TEM and OM, respectively. A graphene suspension, size-fractionated by sedimentation, was used for DLS study. This was characterized in standard quartz cuvette cells without any need for sample deposition.

TEM was initially carried out to determine the lateral dimension and size distribution of the flakes. Owing to the precision of the TEM measurement, the lateral size distribution was then used as a benchmark for further experiments. Flake thickness was estimated by the Mean Grey Value Ratio (MGVR) method which is based on normalised TEM Bright Field image contrast [4]. Direct imaging of folded graphene edges and low-loss electron energy loss spectroscopy (EELS) were used to provide complementary information for MGVR measurements. In addition, an image technique was developed so that the flake lateral size distribution could be obtained rapidly via quantification of several OM images. To develop an in-situ characterisation technique, DLS was performed to obtain the lateral size distribution of dispersed flakes. Although no precise or reliable parameters for graphene flakes dispersed in IPA exist in the literature, approximate optical parameters were used to perform the DLS analysis.

TEM measurements were conducted using an FEI Titan^3 Themis 300 S/TEM operated at 80 kV, which is below the threshold for knock-on damage [5]. TEM magnifications of 55,000 x and 295,000 x were used for the development of MGVR and folded edge methods respectively. EELS measurements were recorded in diffraction mode from an area of ca. 100nm in diameter. Use of STEM could improve the spatial resolution of such measurements in the future. An Olympus BX51 series reflection light microscope was employed for OM, using a 100x objective lens (N.A. = 0.95) and 163 ms exposure time; white balance and RGB ratio were optimised by the pre-installed AxioVision software. A Malvern Zetasizer Nano ZS series was employed for DLS using standard quartz cuvette cells. The sample was equilibrated to *25°C* for *120* seconds prior each measurement. The viscosity of IPA was set to be *2.32 cP* at *25°C*. Using a *633 nm* laser and by operating in backscatter mode (*173°* scattering angle), the particle size can be detected using optimized beam-positioning. Different material refractive indices (RI) and absorption coefficients (α) between graphene and graphite were used to derive number particle distributions (PSDs). Parameters of: (1) 1nm thick graphene ($RI_{graphene} = 2.225; \alpha_{1nm} = 3\%$) and (2) 3nm thick graphite ($RI_{graphite} = 1.942; \alpha_{3nm} = 10\%$) were employed [6]. Each of the PSDs were the average of 3 measurements. Image data was processed by Fiji or GMS 3. The Scipy package and OriginPro were used for data analysis and visualisation.

3. **Results**

3.1 Imaging techniques

Graphene flakes are often aggregated or partially folded, complicating images and making them hard to quantify. Example TEM images are shown in fig. 1 (a) and (b); lateral flake size measurements are illustrated by the yellow lines (Feret diameter). To further quantify TEM images, the Mean Grey Value Ratio (MGVR) method was used. Precise flake thicknesses were first estimated by (002) lattice imaging at folded flake edges and corresponding MGVR values were determined by $MGVR = \mu_s/\mu_v$, where μ_s is the MGV of the flake and μ_v is the MGV of the neighbouring vacuum region (figure 2(a)). A linear correlation between the MGVR and the flake thickness is shown in figure 2 (b). Using the following empirical expression, the number of graphene layers (n) can be estimated by:

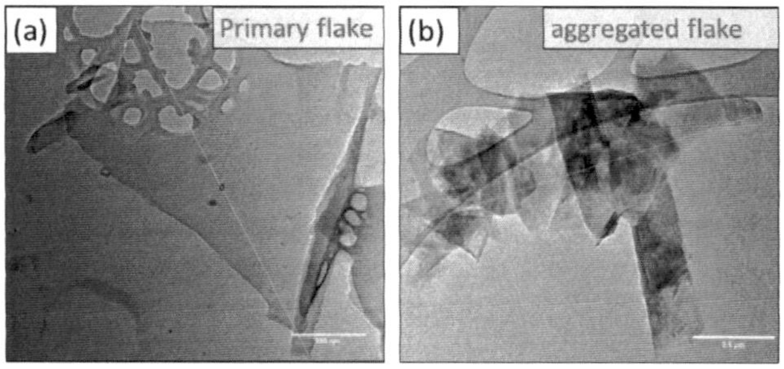

Figure 1. Bright field TEM image of graphene (a) primary flake and (b) aggregated flake. Size measurements are illustrated by yellow lines (Feret diameter)

$$n = MGVR - (1.0358 \pm 0.017)/(-0.012 \pm 0.001) \quad (1)$$

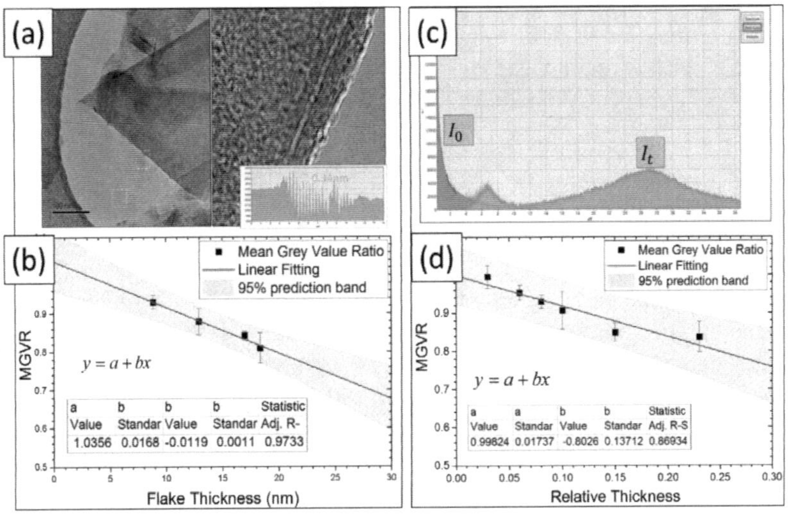

Figure 2. Flake thickness estimation by TEM. (a) and (b): comparison folded edge and MGVR method. (c) Low-Loss EELS spectrum and (d) comparison of relative thickness derived by MGVR and low-loss EELS

This relationship was further supported by calculating the relative thickness (t/λ) from the low-loss EELS spectrum. The relative thickness was calculated using $(t/\lambda) = \log(I_t/I_0)$. t and λ are the absolute thickness of sample and the electron inelastic mean free path in nm; I_t and I_0 are the integrated areas under the entire low-loss spectrum and the zero-loss peak respectively (figure 2(c)). Correlation between the MGVR and the relative thickness is shown in figure 2(d).

The distribution of lateral flake sizes was obtained by analysis of TEM images and showed mean sizes of $\langle D \rangle = 0.619 \pm 0.32 \mu m$ and $\langle D \rangle = 1.236 \pm 0.43 \mu m$ for primary and aggregated flakes, respectively (figure 3(c)). A similar bimodal distribution was obtained from OM. Using the variation of RGB contrast versus flake thickness, flakes of interest can be identified by a computer algorithm (figures. 3(a) and (b)). The size distribution was then derived via analysis of *6572* thin graphene flakes (blue spots) from several OM images, giving a mean *lateral size of* $\langle D \rangle = 0.776 \pm 0.345 \mu m$. As seen in Figure 3(c), the distribution was fitted by two Gaussians, exhibiting only *0.9 %* and *0.5 %* mean size differences from the distribution of primary and aggregated flakes derived by TEM. Owing to the increased sampling of flakes, a smoother distribution was obtained from OM as compared to TEM.

To gain insight into flake thicknesses, Eq.1 was used to estimate the thicknesses of primary and aggregated flakes as *37 ± 13 layers (MGVR = 0.884 ± 0.097)* and *76 ± 29 layers (MGVR=0.725 ± 0.191)* respectively. Although flake thicknesses are difficult

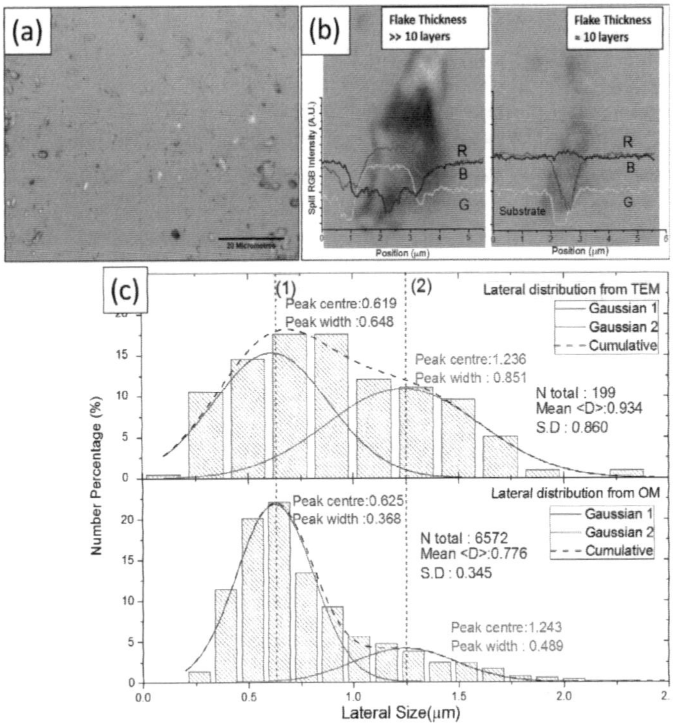

Figure 3. The lateral size distribution obtained by image analysis techniques: (a) an example of OM image. (b) Split RGB images of thick and thin graphene flakes. (c) Comparison of lateral size distributions obtained by TEM (above) and OM (below)

to determine precisely from the MGVR method, it is evident that flake aggregation is preferable vertically rather than horizontally. Determination of flake thickness by OM is even more difficult; even though it has been reported that flakes >>10 layers appear yellow/brown and flakes ~10 layers appear blue/purple on such SiO_2/Si substrates [7], inadequate colour resolution makes thickness estimation by OM highly imprecise.

3.2 Dynamic Light Scattering (DLS)

In DLS, the lateral flake PSD is highly influenced by sample concentration. Thus, the suspensions were size-fractionated to minimize uncertainty. In figure 4, the peak position of number PSD (Xc) from DLS was plotted versus the mean lateral flake diameter measured by OM (labelled as $< D >$). The Xc values using selected optical parameters are close to each other and scale linearly with $< D >$ on the log-log plot. This means the Xc is correlated to < D > by a power exponent. As in reference [8], the data was fitted using $Xc = a < D >^{b}$, where the exponent $b = 1.594 \pm 0.12$ and $a = 0.018 \pm 0.015$. Using the data in figure 4, we can write:

$$< D > = (12.433 \pm 4.33) Xc^{(0.627 \pm 0.05)} \quad (2)$$

For each of the size fractionated samples, DLS and OM results deviate by less than *22%* for flake sizes $<$ *1000 (nm)*, with results from DLS being consistently larger. The expression can provide an alternative approach for determining graphene lateral size distribution from DLS number PSD with a similar precision to direct imaging techniques in the sub-micrometre region. However, weaknesses of this approach arise from the effects of sample concentration, whereby the signal is too low to detect when the absorbance is $< 0.001(m^{-1})$. Also, though the deviations between DLS and OM are small in sub-micron region, it increases significantly for flake sizes $>$ *1000 nm*.

4. Conclusions:

We present a simple and rapid method to estimate the lateral size distribution of flakes of solution-processed graphene which is highly important for its applications. Using imaging and image analysis, good correspondence was found between precise measurements made by TEM and automated image analysis of OM images. Flake thickness was also estimated by MGVR of TEM images, which is difficult to achieve by OM. Results from DLS were then compared to the OM measurements, suggesting that DLS could provide a rapid screening for graphene lateral size distribution.

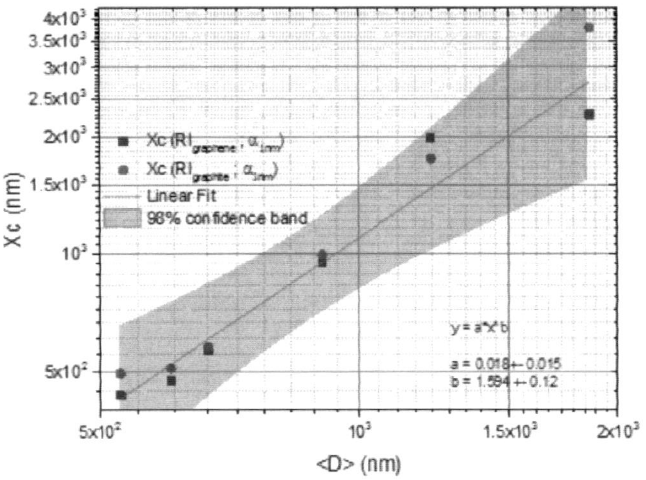

Figure 4. Plot of number PSD peak centre (Xc) versus flake mean lateral size < D >. The red line is the fitted power law dependence of Xc with < D >

References

[1] N. V. Kozhemyakina 2010 *Adv. Mater.*, vol. 22, no. 48, pp. 5483–5487.
[2] K. Parvez 2013 *ACS Nano*, vol. 7, no. 4, pp. 3598–606.
[3] B. Terrés 2016 *Nat. Commun.*, vol. 7, p. 11528.
[4] S. Rubino 2016. *Microsc. Microanal.*, vol. 22, no. 01, pp. 250–256.
[5] F. Banhart 1999 *Reports Prog. Phys.*, vol. 62, no. 8, pp. 1181–1221.
[6] M. N. Polyanskiy *Refractive index database*.
[7] Z. H. Ni 2007 *Nano Lett.*, vol. 7, no. 9, pp. 2758–63.
[8] M. Lotya 2013 *Nanotechnology*, vol. 24, no. 26, p. 265703.

IOP Conf. Series: Journal of Physics: Conf. Series **902** (2017) 012027 doi:10.1088/1742-6596/902/1/012027

Eco-Friendly Synthesis and Characterization of Reduced Graphene Oxide

D. Ickecan[1], R Zan[2], S. Nezir[1]

[1]Department of Physics, Kırıkkale University, Kırıkkale, Turkey
[2]Department of Physics, Omer Halisdemir University, Nigde, Turkey

recep.zan@ohu.edu.tr

Abstract. Graphene is a single sheet of sp^2 bonded carbon having a two-dimensional (2D) layer. It has remarkable electronic, mechanical and thermal properties. In this paper, the graphene oxide (GO) was reduced by reducing chemicals such as ascorbic acid and hydrazine and then characterized by transmission electron microscopy (TEM), Raman spectroscopy and Fourier transform infrared spectroscopy. TEM results of the chemically reduced graphene were showed that the structure consists of a mixture of single and few layers of reduced graphene oxide (rGO).

1. Introduction

Graphene is a hexagonal lattice structure of one atom thick layer and includes σ bond between sp^2 hybridized carbon atoms. Graphene has properties such as outstanding surface area to mass ratio (2,630 m^2/g), high thermal conductivity (5,000 W/mK), optical transparency (~97.7 %) and high electron mobility ($2,5\times10^5$ cm^2/Vs) [1-3]. The unique properties of graphene can be employed in a range of applications such as electronic, composite material, batteries and sensor technology. Due to the high demand for graphene, several fabrication methods such as epitaxial growth (SiC) [4], chemical reduction [5], chemical vapour deposition (CVD) [6] and liquid phase exfoliation [7], have been developed in recent years. Among the four methods, chemical synthesis is considered the best since it is easier, inexpensive and yields large quantities. The researchers have applied different types of reducing agents in order to convert graphene oxide to graphene and then confirmed that hydrazine and ascorbic acid had the ability to remove oxygen functionalities of GO [8,9].While the hydrazine is highly toxic, the ascorbic acid is an eco-friendly and non-toxic reducing agent.

2. Experimental

Materials: Natural graphite power, concentrated sulfuric acid (H_2SO_4, 98%), Potassium permanganate ($KMnO_4$), Hydrogen peroxide (H_2O_2,30%), L-(+)- Ascorbic acid, L-trypotophan, Phosphorus pentoxide, hydrazinehyrate, Potassium persulfate aqueous solution were all analytical grade and purchased from Sigma Aldrich.
Preparation: The technique employed in this study is a two-step process. In the first step, graphene oxide is obtained by oxidation of graphite powder and the following step includes the reduction of graphene oxide. Graphene oxide was prepared by the Hummers method [10] (HGO) and modified

Hummers method (MGO) [11]. Subsequently, graphene was synthesized by using reductant chemicals such as hydrazine hydrate [12] (HIGO) and ascorbic acid [13] (VIGO).

Characterization: Graphene oxide, MGO, HIGO and VIGO sample were freeze-dried and then grounded into power. The CTEM images was obtained using Model FEI Tecnai G2 (TWIN) the range from 20-200 kV. The Raman spectra were recorded from 100 to 3500 cm^{-1} on Senterra Bruker Raman Microprobe using a 532 nm/20nW Nd oil laser. The FTIR spectra of graphene oxide, HIGO and VIGO samples were recorded over the range 4000-400 cm^{-1} on a Vertex 70v FTIR spectrometer.

3. Results and Discussion

The surface morphology and crystallinity of the GO and rGO were analyzed using CTEM and Selected Area Electron Diffraction (SAED) pattern as shown in Figure 1. Both GO and rGO showed transparent, surface rough or wrinkled structure under the CTEM. The CTEM result of the samples HGO and MGO shows the sheet like morphology comprised of many layers of graphene oxide. Researchers defined ultrathin silk veil structure with fold and scroll on its edges [14].

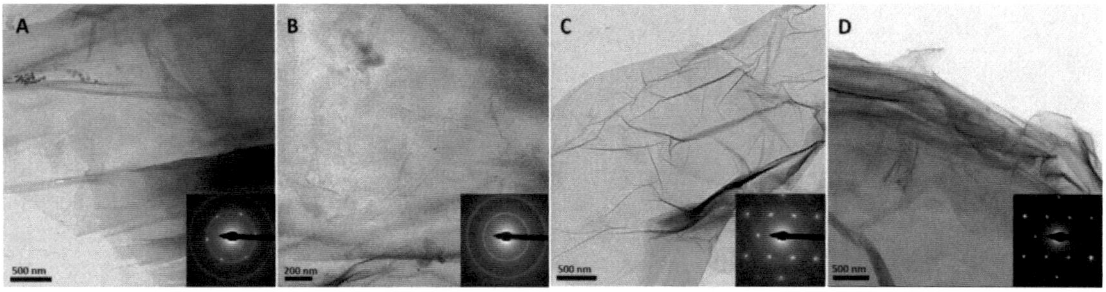

Figure 1. a) CTEM images of HGO, b) CTEM images of MGO, c) CTEM images of HIGO, d) CTEM images of VIGO.

The structure of graphene oxide (GO) is different from reduced graphene oxide (rGO) in SAED analysis. It was observed that the SAED pattern of the rGO has clear diffraction spots with a six-fold pattern that is compatible with the hexagonal lattice. Moreover, for single layer rGO, the intensity of the outer hexagon spot is almost similar to or less than that of the inner one. According to the SAED analysis, the structure consists of a mixture of single and few layers of rGO (VIGO).

Raman spectroscopy is an efficient technique to determine the number of graphene layers and the change of crystal structure of GO and rGO. There are two main bands, the D band and the G band together with weak 2D (also called the G′) band. The D band is related to disorders and defects in graphene. The G band is Raman active for sp^2 hybridized carbon-based material. The 2D (the G′) band is used to determine whether the graphene is single or few layered. Raman spectrum of graphite, GO and rGO are shown in figure 2 and 3.

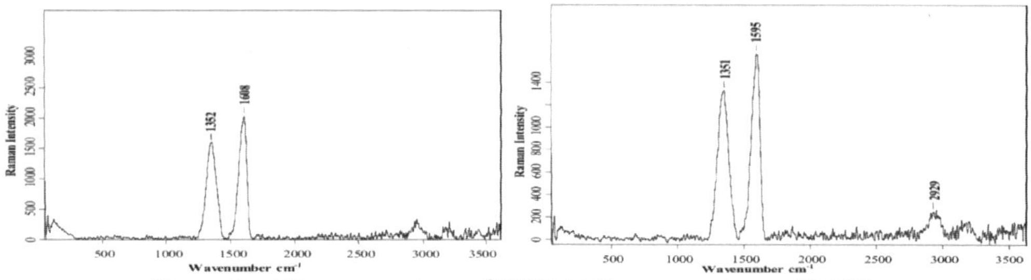

Figure 2. a) Raman spectrum of HGO, b) Raman spectrum of MGO.

The two main bands in the graphite spectrum are known as the G band at ~1582 cm^{-1} and the 2D band at ~2685 cm^{-1}. On the other hand, The Raman spectra of graphene oxide (HGO and MGO) displays a D band at ~1350 cm^{-1} and a broad G band at ~1580 cm^{-1}. The D band of the Raman spectrum of graphene oxide is prominent. This results in the degradation of the in-plane sp^2 domain in graphene oxide owing to the extensive oxidation.

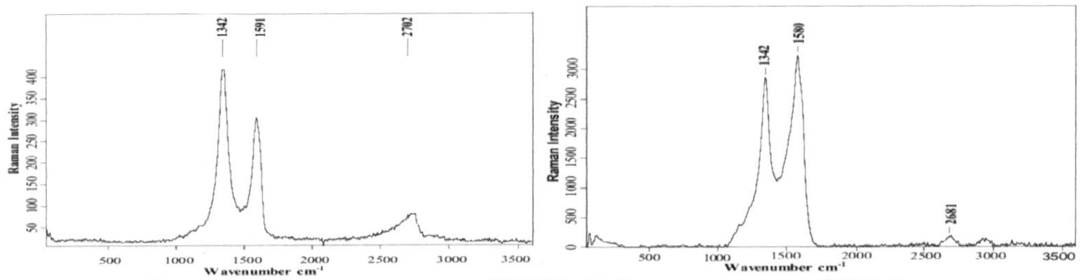

Figure 3. a) Raman spectrum of HIGO, b) Raman spectrum of VIGO.

Raman spectra of the VIGO are observed in the D band at approximately 1340 cm^{-1} and the G band at 1580 cm^{-1} together with a weak 2D band around 2700 cm^{-1} [15]. Structurally, the D band and G band of the rGO are sharper and more intense as against those of graphene oxide. Moreover, the intensity ratio of D band and G band (I_D/I_G) indicates the oxidation degree and the size of sp^2 ring clusters in a sp^3/sp^2 hybrid network of carbon atoms. The (I_D/I_G) intensity ration chemically reduced graphene is 1,37 which is larger than that of graphene oxide 0,80. This result suggests that more sp^2 domains are formed during the reduction of graphene oxide.

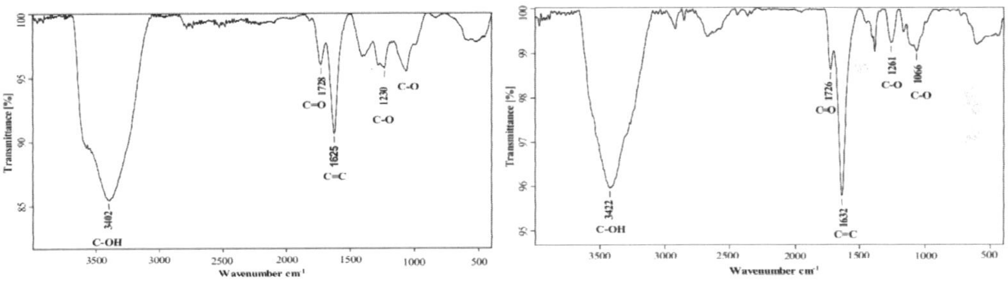

Figure 4. a) FTIR spectra of HGO, b) FTIR spectra of MGO.

To investigate the different types of functional groups formed in the GO and rGO, FT-IR spectroscopy was used. The FT-IR spectra of graphene oxide (HGO and MGO) are shown in figure 4. Firstly, the broad peak at 3000-3500 cm^{-1} originates from stretching vibrations of -OH group, and the peak observed at 1700-1750 cm^{-1} stretching vibrations of carboxyl peaks (C-O). Another aromatic C=C peak appears between the peak at 1600-1650 cm^{-1} and at 1000-1280 cm^{-1} stretching vibrations of epoxy C-O groups. Finally, the peak showed alkoxy stretching vibrating between the peak at 1040-1170 cm^{-1}.

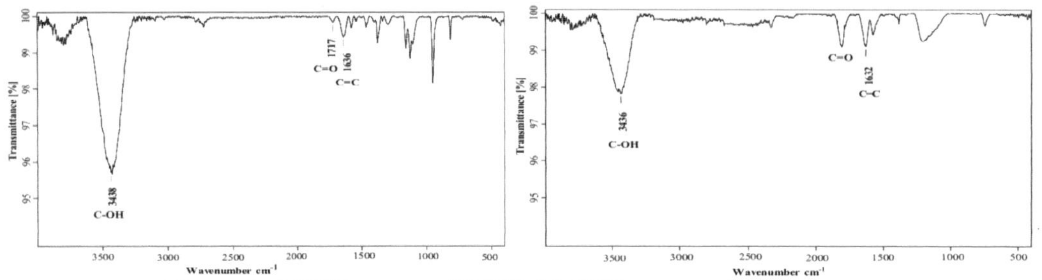

Figure 5. a) FTIR spectra of HIGO, b) FTIR spectra of VIGO.

The FT-IR spectra of rGO in figure 5 shows three peaks, the broad and intense peak at ~3400 cm^{-1} arising from–OH stretching vibration, a peak at ~1720 cm^{-1} due to C=O stretching vibrations of carboxylic groups and sharp C=C band within the range of 1600 cm^{-1}. As can be seen in Figure 5, the absorption peaks containing oxygen such as, epoxy and alkoxy vibration peaks disappeared in HIGO and VIGO. This result confirms that the atomic frame of sp^2 carbon is formed in reduced by graphene oxide.

4. Conclusions

The reduction of graphene oxide by ascorbic acid formed sp^2 structures and was found to increase regular structures in comparison to hydrazine. The chemical reduction of graphene oxide by ascorbic acid is found to be a very promising technique for large scale production of reduced graphene (rGO). The ascorbic acid reduction technique does not contain toxic agents, which provides eco-friendly and high quality graphene when compared to other known techniques. The use of hydrazine hydrate as the reductant has also been revealed not to be biocompatible. Consequently, excellent chemical reduction of graphene oxide by ascorbic acid can be utilized for tissue engineering and other biomedical applications in the future.

References

[1] W. Cai, Y. Zhu, X. Li, R. D. Piner and R. S. Ruoff, Appl. Phys. Lett., 2009, 95(12), 123-115.
[2] X. Li, Y. Zhu, W. Cai, M. Borysiak, B. Han, D. Chen, et al., Nano Lett., 2009, 9(12), 4359–4363.
[3] V. Chabot, D. Higgins, A. Yu, X. Xiao, et al., Energy Environ. Sci., 2014, 7(5),1564–1596.
[4] T. Ohta, A. Bostwick, T. Seyller, K. Horn and E. Rotenberg, Science, 2006, 313(5789), 951–954.
[5] Eda, G., G. Fanchini, and M. Chhowalla, Nature Nanotechnology, 2008. 3(5): p. 270-274.
[6] Reina, X. Jia, J. Ho, D. Nezich, H. Son, V. Bulovic, M. S. et al., Nano Letters 9 (2008) 30.
[7] Y. Hernandez, V. Nicolosi, M. Lotya, F. M. Blighe, et al.,Nature Nanotechnology, 2008, 3(9),563–568.
[8] Park S, An J, Potts JR, Velamakanni A, Murali S, Ruoff RS., Carbon. 2011, 49(9), 3019–23.
[9] Fernández-Merino MJ, Guardia L, Paredes JI, Villar-Rodil S,et al.,J Chem Phys. 2010,114,6426–32.
[10] W.S. Hummers and R.E. Offeman, Journal American Chem. Soc., 1958, 80, 1339.
[11] S. Liu S, et al., ACS Nano 2011,5, 6971–6980.
[12] S. Stankovich S, et al., Carbon 2007, 45, 1558–1565.
[13] J. Gao, et al., Chemisty of Materials, 2010, 22, 2213-2218.
[14] L. Tang, Y. Wang, Y. Li, H. Feng, J. Lu and J. Li, Adv. Funct. Mater., 2009, 19(17), 2782
[15] L. Xiao, Springer, 2015, 29.

Acknowledgements

This research was supported by Kırıkkale University under the BAP contact number 2016/47.

IOP Conf. Series: Journal of Physics: Conf. Series **902** (2017) 012028 doi:10.1088/1742-6596/902/1/012028

Imaging Two Dimensional Materials and their Heterostructures

R Zan[1], Q M Ramasse[2], R Jalil[3], J-S Tu[4], U Bangert[5] and K S Novoselov[4]

[1]Department of Physics, Omer Halisdemir University, Nigde, Turkey
[2]SuperSTEM, STFC Daresbury Laboratory, Warrington, UK
[3]Department of Physics, University of Engineering & Technology, Lahore, Pakistan
[4]School of Physics and Astronomy, The University of Manchester, Manchester, UK
[5]Department of Physics and Energy, University of Limerick, Ireland

recep.zan@ohu.edu.tr

Abstract. Stacking different two-dimensional (2D) atomic layers is a feasible approach to create unique multilayered van der Waals heterostructures with desired properties. 2D materials, graphene, hexagonal boron nitride (h-BN), molybdenum disulphate (MoS_2) and graphene based van der Waals heterostructures, such as h-BN/graphene and MoS_2/graphene have been investigated by means of Scanning Transmission Electron Microscopy (STEM).

1. Introduction

In recent years, two-dimensional (2D) materials including graphene, hexagonal boron nitride (h-BN), and transition metal dichalcogenides (TMDs) have attracted great attention for applications in fields. In particular, the combination of these layered materials to obtain van der Waals heterostructures is of remarkable interest [1].

Single layer hexagonal boron nitride (h-BN) in contrast to a gapless semiconductor graphene has a wide band gap of ~5.9 eV, which arguably makes it the thinnest possible insulator and therefore a perfect candidate for incorporation into heterostructures [2]. One of the most studied heterostructure type is known to be graphene/h-BN. More recently, transition-metal dichalcogenides such as molybdenum disulphide (MoS_2) and tungsten disulphide (WS_2) have also been under heavy investigation as a new class of materials with remarkable properties. By implementing these single layer materials into graphene based devices, the device performance can be enhanced [3]. Single layer MoS_2 is a direct gap semiconductor with 1.8 eV that is different from its bulk form and it has an indirect band gap of 1.2 eV [4]. Although electrical, magnetic and mechanical studies of 2D materials and their combination to obtain heterostructures are established, there is a lack of direct atomic visualisation of these systems, thus limiting our understanding of these highly exciting materials [5,6].

In this paper, we present a (S)TEM study of graphene/h-BN and graphene/MoS_2 heterostrusctures with an emphasis on their lateral interfaces where heterostructures meet the graphene layer.

2. Experimental

Graphene, h-BN and MoS_2 were prepared via mechanical exfoliation technique from their bulk. Graphene, h-BN and MoS_2 layer numbers were identified via light microscopy and Raman spectroscopy. Following the fabrication and characterization of the 2D materials, h-BN layer was transferred onto graphene layer via wet chemistry approach to obtain graphene/h-BN heterostructure by using mask aligner. The same procedure was implemented to fabricate graphene/MoS_2 heterostructure.

Screening and diffraction imaging of the samples were carried out in a stationary transmission electron microscopy mode (TEM) in a FEI Tecnai F30. The detailed electron microscopy studies were performed in an aberration corrected dedicated STEM (Daresbury SuperSTEM) at low primary beam energy (60 kV) with angstrom-sized probe. This acceleration voltage is known to be lower than the threshold of knock-on damage for most of the sensitive 2D materials. Atomic sites of carbon, boron, nitrogen, molybdenum and sulphur atoms and lateral interface of the heterostructures were directly revealed in medium angle annular dark field (MAADF) imaging and high angle annular dark field (HAADF) imaging mode by their approximate Z^2 dependence and electron energy loss spectroscopy (EELS). All presented images in this paper are raw (unprocessed).

3. Results and Discussion

Figure 1 shows low magnification TEM image and diffraction patterns of the graphene/h-BN heterostructure. In the low magnification TEM image in figure1(a) four holes in lacey carbon film on the TEM grid are fully covered with single layer graphene and the single layer h-BN covers almost two holes, the right side of the figure1(a). As a result of this, we managed to investigate the junction between graphene and the heterostructure. As can be seen in the same figure, h-BN covering the region of graphene (heterostructure) can reveal itself with bubbles, darker and thicker regions due to the transfer of h-BN onto graphene as the transfer process involves chemicals that are expected to contaminate the structure. The transfer process may also lead to strain in the heterostructure. The diffraction patterns of the graphene layers and the heterostructure that were taken from the colored boxes in figure1(a) are presented in figure1(b) and figure1(c) respectively. The diffraction pattern of graphene confirms that the graphene is purely single layer as shown in figure1(b). Since graphene (2.46 Å) and h-BN (2.52 Å) have similar lattice constants, they both appear very similar in the diffraction pattern as shown in figure1(c). There are 12 spots in the inner (and outer) set of spot, 6 of which belong to graphene and the other 6 to h-BN layer.

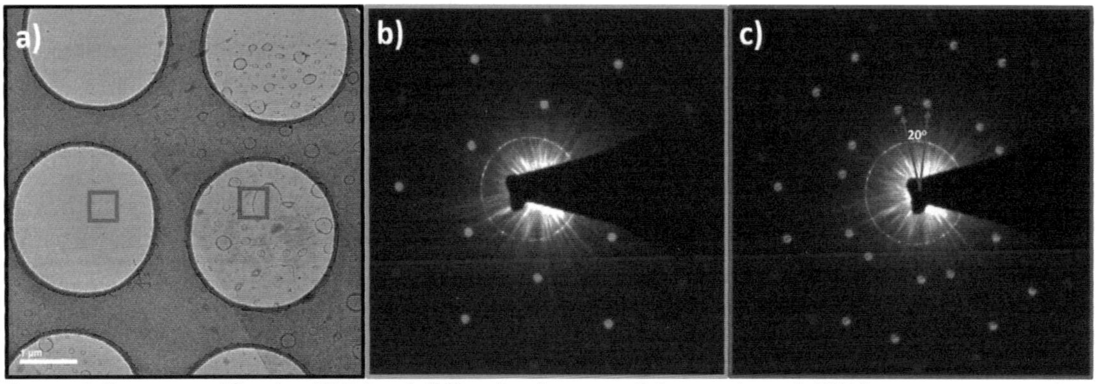

Figure 1. Low magnification TEM and diffraction images of the graphene/h-BN heterostructure, taken at 300kV.

It is found that the angle between the closest spots which are graphene and h-BN spots is measured to be about 20°. This angle also represents the rotation between the layers with respect to each other. As a result, the overlapping graphene and h-BN layer to create heterostructure can cause Moire patterns which have larger periodicities than graphene and h-BN depending on the rotation angle of the layers. Moire pattern is observed even in perfect matching due to the 1.8% difference in the lattice constant between graphene and h-BN [7,8].

Figure 2 shows STEM-MAADF images of the graphene/h-BN heterostructure as taken at 60kV acceleration voltage. The heterostructure and graphene junction is shown in figure2(a). However, the lateral interface is not clearly visible due to the hydrocarbon contamination over the junction. However, the graphene side of the junction can be seen in figure2(b) as a magnified image of the blue box in figure2(a). Similarly, a closer look at the red box in figure2(a) shows graphene/h-BN heterostructure that reveals itself via Moire pattern that has different periodicity than graphene or h-BN (figure2c). In order to avoid contamination over the junction and to be able to observe the lateral interface, beam shower (about 10mins) was applied to the region that was to be imaged. As a result of this, as can be seen in figure2(d), a clear transition from graphene to heterostructure was observed. However, while imaging the region, we observed knock on damage after few minutes of beam scanning on the h-BN layer at the interface. This is due to the weak bonding at the edge compared to the basal plane as damaging wasn't noticed in the basal plane. On the other hand, much faster damaging of h-BN compared to graphene/h-BN heterostructure was observed in our previous experiments (only h-BN imaging) [9]. Therefore, we can conclude that graphene serves as a shield for h-BN against radiation damage. Additionally, graphene damaging was not encountered during the experiments, either.

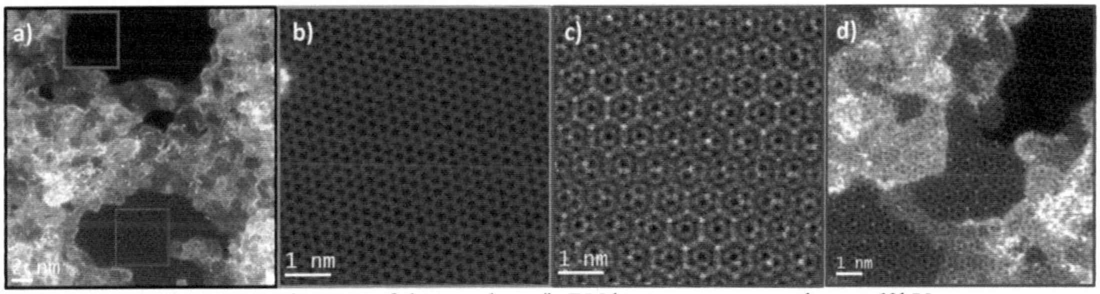

Figure 2. STEM-MAADF images of the graphene/h-BN heterostructure, taken at 60kV.

Similar to graphene/h-BN heterostructure, graphene/MoS_2 heterostructure was prepared by transferring single layer MoS_2 onto single layer graphene. STEM-HAADF images of graphene/MoS_2 heterostructure are illustrated in figure 3. A STEM-HAADF image of the heterostructure and graphene junction is shown in figure3(a) where the dark region represents graphene and the bright side is the heterostructure. Due to the large difference between atomic numbers of graphene (Z_C=6) and MoS_2 (Z_{MO}=42 and Z_{2S}=2x16), the contribution of the graphene to the atomic contrast in the HAADF images (approximate Z^2 dependence) is minimal and this effect is clearly visible in figure3(a). The FFT of the heterostructure that is presented as inset figure3(a) shows the angle between graphene and MoS_2 spots to be 24°, which means graphene and MoS_2 were rotated 24° with respect to each other. Additionally, in contrast to graphene/h-BN heterostructure, Moire pattern wasn't observed in graphene/MoS_2 heterostructures due to the atomic number difference between the layers as MoS_2 contrast dominates the image. The red rectangular in figure3(a) shows where EELS map was taken from and also, the spectrum image of the same region are presented in figure3(b) and (c). Both figures represent the same area and the only difference was the contrast as it was intentionally reversed to make heterostructure (figure3b) and graphene (figure3c) visible. The EEL spectra of the heterostructure side, yellow rectangular in figure3(b) and graphene side, green rectangular in figure3(c) were presented inset figure3(d). Indeed, EEL spectra have proven the presence of graphene and heterostructure. During EELS measurement which was performed in more than a few minutes, we once more observed damage in MoS_2 at the lateral interface similar to the graphene/h-BN heterostructure and not in the basal plane. As a result, it can be inferred that graphene protects the MoS_2 layer from the radiation damage as well [5].

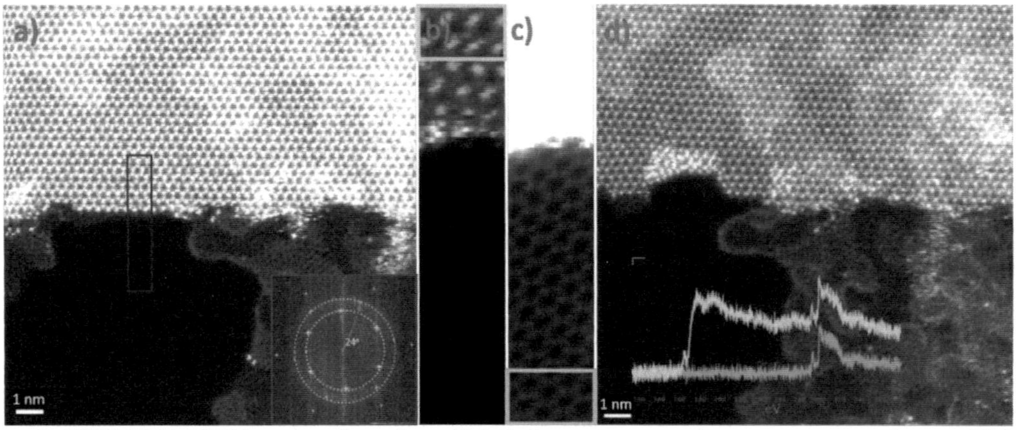

Figure 3. STEM-HAADF images of graphene//MoS_2 heterostructure heterostructure, taken at 60kV.

4. Conclusions

In conclusion, graphene/h-BN and graphene/MoS_2 heterostructures were studied via TEM, diffraction and STEM -MAADF and -HAADF imaging. Moire pattern was observed due to the rotation between the graphene and h-BN layers. However, this wasn't found in graphene/MoS_2 heterostructure despite the existing rotation between the layers because of the huge atomic number difference between graphene and MoS_2. The lateral interface of graphene and the heterostructures was also defined and it was noticed that both h-BN and MoS_2 were damaged at the interface during prolonged imaging due to the weak bonding between the atoms at the edges of the layers as the damage was seen neither in the basal plane of the h-BN nor in the that of the MoS_2. Additionally, it was revealed that graphene serves as a protecting layer from the radiation damage for h-BN and MoS_2 layer.

References

[1] K. S. Novoselov, et al., PNAS, 102, 30, (2005).
[2] C. R. Dean, et al., Nature Nano, 5, 722, (2010).
[3] L. Britnell, et al., Science, 335, 947, (2012).
[4] Q.H. Wang, et al., Nature Nano., 7, 699, (2012).
[5] R. Zan, et al., ACS Nano, 7, 10167, (2013).
[6] S. J. Haigh et al., Nature Materials, 11, 9, (2012).
[7] C. R. Woods, et al., Nature Phys., 10, 451, (2014).
[8] O. Cretu et al., Micron, 72, 21, (2015).
[9] R. Zan, et al. Journal of Microscopy, 244 (2), 152-158, (2011).

Acknowledgements

The authors gratefully acknowledge funding from the Turkish Academy of Sciences under Outstanding Young Scientist Award Programme (TÜBA-GEBİP).

TEM-based Pair Distribution Function study of interatomic distances in C-supported Pt

D T Tran[1,2] and D Ozkaya[1,2]

[1]Johnson Matthey Technology Centre, Blounts Court Road, Sonning Common, Reading RG4 9NH, UK.
[2]ePSIC, I14 Harwell Science and Innovation Campus, Fermi Ave, Didcot OX11 0DE, UK.

trung-dung.tran@diamond.ac.uk

Abstract. The interatomic distances have been characterised for a Pt on Carbon based fuel cell catalyst studied by analysing the atomic pair distribution functions (PDF) obtained from electron diffraction (ED) data taken in a transmission electron microscope (TEM). The experimental PDFs have been compared with atomistic models to examine C-C and Pt-Pt interatomic distances. Further, the models have been refined by reverse Monte-Carlo simulations (RMC) based on the experimental PDF, enabling the investigation of Pt-C interatomic distances. This has demonstrated the existence of an interatomic contact between the Pt and C.

1. Introduction
Most TEM-based studies of Pt on Carbon type fuel cell catalysts only focus on the particles while considering the supporting media as background. However, the Pt-C interaction has an important role on the stability of the system under catalytic conditions, as it affects diffusivity of the Pt on carbon. This in turn causes changes in size, shape, dispersion, and crystallographic ordering of nanoparticles. This interaction needs to be understood at the atomic level. However, it is very challenging to atomically observe Pt particle-carbon interaction by the popular high-resolution (HR)TEM and aberration-corrected scanning (S) TEM due to difficulties of getting a particle edge on in an amorphous structure.

For amorphous/disordered structures, electron diffraction (ED) yields diffuse scattering rather than sharp Bragg peaks, similar to X-ray/neutron diffractions. Subsequent Fourier analysis of the diffraction data gives a new representation of data in real space, called atomic pair distribution function (PDF), which is actually the function of interatomic distances, resulting in the probability densities of finding atomic pairs [1]. Beside X-ray/neutron diffractions, ED data can also be used to achieve PDFs [2]. More quantitative ED-based PDFs can be obtained by appropriate background corrections and scaling using the scattering factors for electrons [3].

Here we have used the same ED based methodology to study Pt nanoparticles on amorphous carbon support. The experimental PDF has been compared with the theoretical PDFs calculated from atomistic models of Pt supported on disordered carbon. These models have been then refined in RMC simulations in order to extract the Pt-C interatomic pair distributions.

2. Methods

2.1. Experimental methods of TEM-based PDFs
The imaging and ED experiments of Pt nanoparticles supported on carbon was performed using the JEOL ARM200CF TEM operating at 200 kV acceleration voltage. The sample was imaged in STEM mode, including bright field (BF) and high angle annular dark field (HAADF) signals. Using the Orius SC200D CCD camera, selected-area ED data was acquired (without any beam stoppers) for the

momentum transfer (Q) range from Q_{min} ~ 2 Å^{-1} to Q_{max} ~ 14 Å^{-1}. The routines to obtain the PDF from the ED data were carried out using the software SUePDF [3].

2.2. Observations and modeling methods

In order to build appropriate models which are as close to reality as possible, HRSTEM images of the C-supported Pt, shown in Figure 1, were examined. The particle predominant sizes are between ~ 2 – 5 nm. The C support, as seen in Figure 1(a, b), exhibits some disordered graphitic structures. The metallic Pt, in contrast, exists in two forms: fcc-crystalline nanoparticles and individual atoms (Figure 1(b, c)). It is noted that, the Pt atomic dispersion may suggest a significant interaction between the metal and the C-support. The particles are generally based on truncated octahedral (TO) symmetry in shape. Their landing direction onto the C support appears to be random. The direction can be the basal planes, the edge planes, or planes in between these, due to the disordered nature of the C-support.

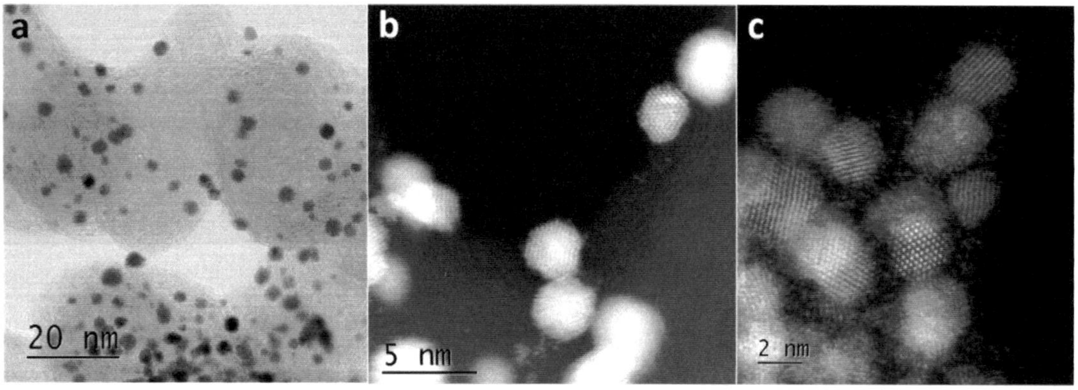

Figure 1. HRSTEM images of C-supported Pt nanoparticles: (a) BF image; (b)HAADF image with graphitic contrast; (c)HAADF image with particle crystallinity and Pt atomic dispersion.

Following the observation of the particles and the C-support, two different atomistic models of C-supported Pt were built. The first model consists of a 481-atom TO-based Pt particle (~2.5 nm) landing toward the basal planes of a 528-atom graphite structure. The second model is larger, having a 1127-atom TO-based Pt particle (~3.5 nm) supported vertically to the edge planes of a 2546-atom disordered graphitic structure. To build the initial Pt-C contact in both models, Pt and C atoms were randomly moved into a range of 2.0-2.4 Å.

These models were used as the initial inputs for RMC simulations [5, 6] where the atomistic structures are refined to fit better with the experimental data. In our RMC simulations, the atoms of both Pt and C were allowed to move and the calculation of theoretical PDFs included the convolution with the termination function [4] caused by the limited experimental Q range. To maintain the physical meaning of interatomic distances, minimum distance constraints were set for Pt-Pt, Pt-C and C-C pairs as 2.55 Å, 2.00 Å and 1.15 Å, respectively.

3. Results and discussion

The ED intensity profile of the C-supported Pt is shown in Figure 2a, which has been background-subtracted and scaled using the electron scattering factors for Pt and C. The inset of Figure 2(a) shows the reconstructed ED ring pattern, cropped to a dimension of 20 Å^{-1}. The corresponding PDF is shown in Figure 2(b) with 6 first peaks labelled as P1-P6. The peak P1 (1.41 Å) is clearly attributed to the 1st-shell C-C distances. The peaks P3 (2.77 Å), P5 (3.87 Å) and P6 (4.86 Å) can be identified for the 1st-shell, the 2nd-shell (lattice constant) and the 3rd-shell Pt-Pt distances, respectively, of the fcc structure. The minor peaks P2 (2.28 Å) and P4 (3.37 Å) are less clear to identify because they are more sensitive to the termination ripples [4]. However, the range ~ 2 – 2.4 Å where P2 is located may contain the signal of interatomic spacing between Pt and C atoms. Models of C-supported Pt, allowing extraction of Pt-Pt,

C-C and Pt-C partial PDFs (free from termination ripples) will then be necessary to interpret all of the PDF peaks.

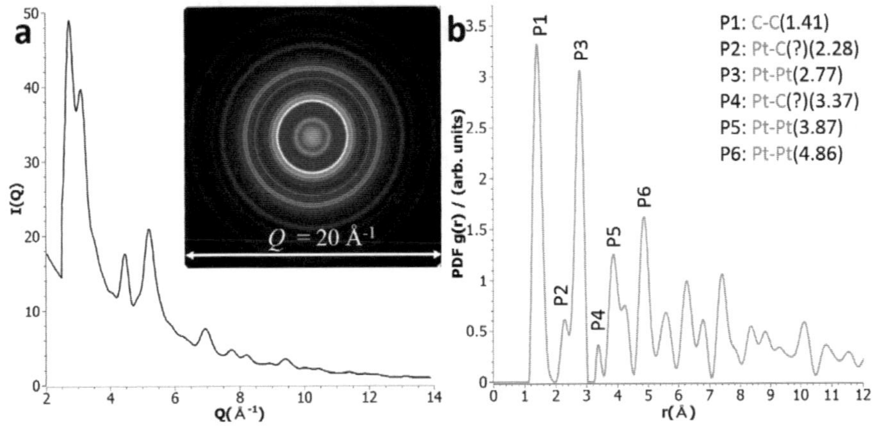

Figure 2. (a) Background-subtracted and normalized ED intensity profile and the corresponding reconstructed ring pattern (inset); (b) The corresponding PDF with peak identification (inset) for P1-P6.

Figure 3 shows the theoretical PDFs of basal-plane supported (Figure 3(a)) and edge-plane supported (Figure 3(b)) models compared with the experimental PDF, before and after the RMC simulations. It can be seen that most of the experimental PDF peaks are correlated with the theoretical ones, confirming the reliability of both the experimental data and the models. However, the initial basal-plane graphite supported model has lower fitness compared with the edge-plane supported one which is based on a disordered carbon structure, due to the disordered nature of the actual C-support (see Figure 1). The fitness of both models has been improved after the RMC refinements, as shown in the bottom panels of Figure3a&b, together with the corresponding RMC-refined atomistic visualizations (insets).

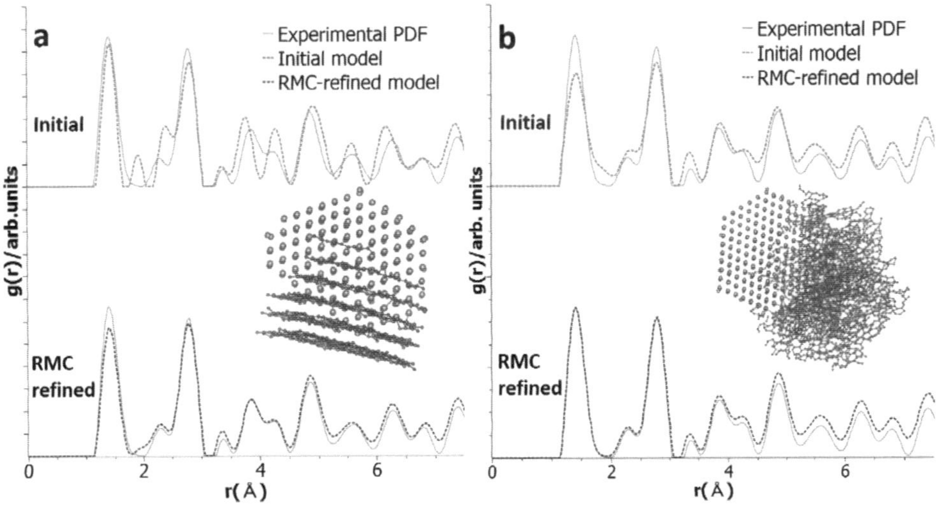

Figure 3. Experimental PDF (orange-solid) compared with theoretical PDFs of initial models (purple-dot) and RMC-refined models (blue-dot), including model visualizations (inset) of Pt (blue) and C (brown) atoms: (a) Basal-plane supported model; (b) Edge-plane supported model.

The Pt-Pt, C-C and Pt-C partial PDFs are shown in Figure 4, exhibiting how the different interatomic distance distributions have been refined after the RMC simulations. The fcc pattern was maintained in the refined Pt-Pt PDFs, consistent with the HRSTEM observation. The C-C PDF of the basal-supported model has a major change from graphite to a disordered graphitic structure. The peaks emerging at ~2.25 Å in the refined Pt-C PDFs of both models clearly indicate a Pt-C interatomic contact between the metal and the support. It is noted that, the Pt-C interatomic peak is more pronounced in the basal-plane supported model compared with the edge-plane supported one, consistent with the fact that the basal planes have more C atoms per unit area than the edge planes do. The strongest peaks in the refined Pt-C PDFs, located at the same position with the first Pt-Pt peak (~2.77 Å), are explained as the indistinguishability between Pt and C atoms during the RMC simulations in the region around 2.77 Å. These peaks correlate with the intensity decrease in the refined first Pt-Pt peaks and the shifts of the refined C-C second peaks, observed in both models.

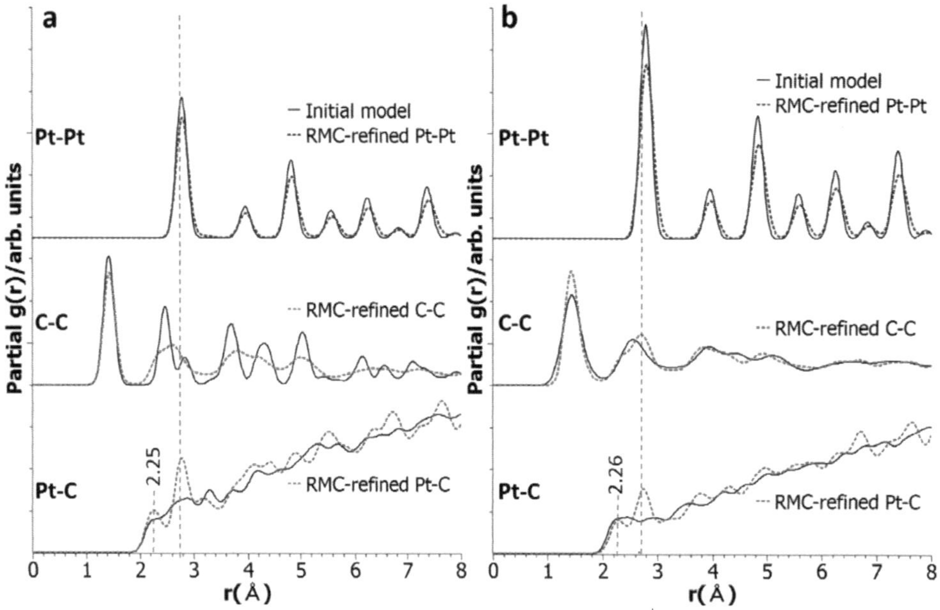

Figure 4. Pt-Pt (blue-dot), C-C (red-dot) and Pt-C (green-dot) partial PDFs extracted from the RMC-refined models compared with the initial models (black-solid): (a) Basal-plane supported model; (b) Edge-plane supported model.

4. Results and discussion

TEM-based PDF analysis has been carried out on Pt nanoparticles supported on disordered graphitic carbon. Two models of Pt nanoparticles supported on graphite basal planes and edge planes of disordered graphic carbon have been built and refined in RMC simulations to interpret the experimental PDF. A Pt-C interatomic contact, peaking at ~2.25 Å, has been revealed from both RMC-refined models. The work has also presented the interatomic-distance measurements of the Pt nanoparticles and the disordered graphitic C-support.

References

[1] T. Egami and S.J.L. Billinge: Underneath the Bragg-Peaks: Structural Analysis of Complex Materials. (Elsevier Science: Amsterdam, 2002).
[2] D. Ozkaya, W. McBride and D.J.H. Cockayne. *Interface. Sci.* **12** (2004), 321.
[3] D. T. Tran, G. Svensson and C.-W. Tai. *J. Appl. Cryst.* **50** (2017), 304.
[4] P. F. Peterson, E. S.Božin, Th. Proffen and S. J. L. Billinge. *J. Appl. Cryst.* **36** (2003), 53.
[5] R.L. McGreevy and L. Pusztai, *Mol. Simul.* **1** (1988), 359.
[6] L. Zeng, D. T. Tran, C.-W. Tai, G. Svensson and E. Olsson. *Sci. Rep.* **6** (2016), 29679.

IOP Conf. Series: Journal of Physics: Conf. Series **902** (2017) 012030 doi:10.1088/1742-6596/902/1/012030

Effect of TEM grid deposition method on the dispersion of thiol-stabilised silver nanoparticles

Julie A Watts[1], Mike W Fay[2], Andrei N Khlobystov[2,3] and Paul D Brown[1,2]

[1]Department of Mechanical, Materials and Manufacturing Engineering, Faculty of Engineering, University of Nottingham, University Park, Nottingham, NG7 2RD.
[2]Nanoscale and Microscale Research Centre, University of Nottingham, University Park, Nottingham, NG7 2RD.
[3]School of Chemistry, University of Nottingham, University Park, Nottingham, NG7 2RD.

E-mail: julie.watts@nottingham.ac.uk

Abstract. Thiol stabilised Ag nanoparticles (NP) in cyclohexane dispersed onto TEM carbon support grids show evidence for agglomeration, whilst the addition of dodecanethiol to the cyclohexane is found to affect the dispersion of the NP which can show highly ordered superlattice formations with close packing, being attributable to the combined effects of convective deposition (coffee ring effect) and the dodecanethiol surfactant.

1. Introduction

Ag nanoparticles (NP) may be regarded as generic building blocks for a range of nanostructured processes [1] and there is particular interest in the control of their size and size distribution, *e.g.* for optical emission wavelength [2]. Here, we report on the development of Ag NP synthesised using a modified Brust-Schiffrin [3] reduction method, and the effect of dodecanethiol addition on their apparent size distribution, at the sample dispersion stage onto the TEM support grid.

2. Materials and Methods

To a solution of silver nitrate (0.3 mmol) in 30 mL ethanol, a mixture of 50 μL 1-dodecanethiol and 60 μL 11(1*H*-pyrro-1-yl)undecane-1-thiol was added slowly, followed by vigorous stirring for 15 minutes. To this was added dropwise, a solution of sodium borohydride (saturated) in 60 mL of ethanol and the solution again stirred vigorously for a further 2 h. The product was then precipitated by the addition of 250 mL of ethanol and refrigerated at -15°C for 24 h. After filtration through a 0.2 μm PTFE membrane the product was washed with 10 mL toluene, followed by washing with 200 mL acetone and drying under vacuum to produced mixed thiol stabilised Ag NP (Figure 1a). The product was then dispersed in cyclohexane, with and without dodecanethiol (Figure 1b), and drop cast onto holey carbon / Cu TEM sample grids. Samples were observed using a JEOL 2100F TEM, operating at 200kV, and particle size histograms were generated using ImageJ to appraise the effects of dodecanethiol addition at the sample dispersion stage. Whilst prolonged exposure to an intense electron beam can lead to NP growth, such effects were avoided by minimising irradiation time and intensity.

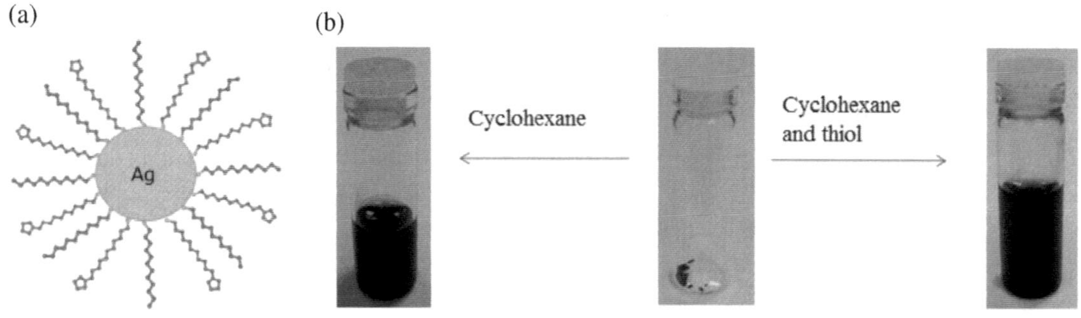

Figure 1. (a) Mixed thiol stabilisation layer around Ag NP; and (b) Ag NP product dissolved in cyclohexane with and without dodecanethiol.

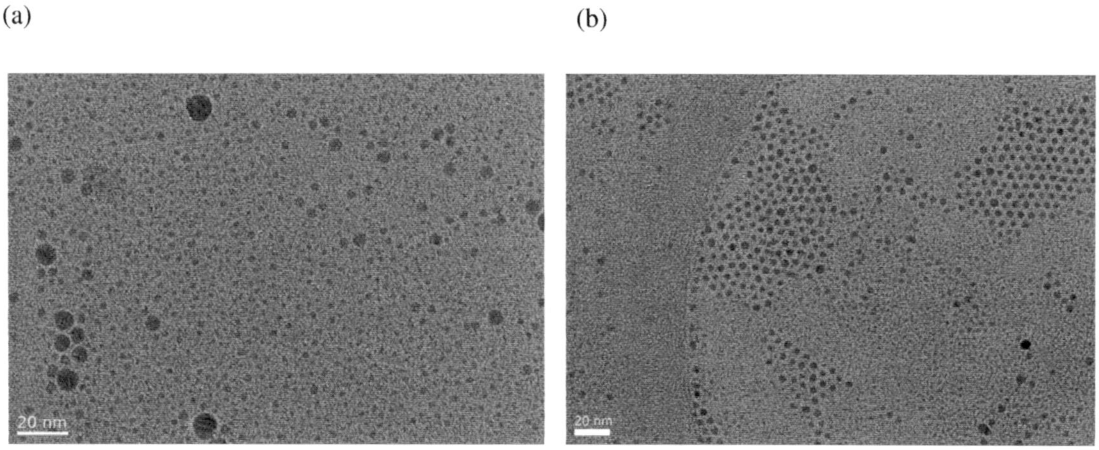

Figure 2. (a) Thiol stabilised Ag NP dispersed from cyclohexane; (b) Thiol stabilised Ag NP dispersed from cyclohexane in the presence of excess dodecanethiol.

3. Results

Figures 2 (a,b) present conventional bright field, diffraction contrast TEM images of thiol stabilised Ag NP dispersed from cyclohexane, and cyclohexane with 30% dodecanethiol (aka dodecanemercaptan DDM), respectively. It is evident that the final processing step has had a significant effect on the NP size distribution as a consequence of the dispersion. After the drying process on the TEM support grid, Ag NP dispersed from cyclohexane showed a bimodal size distribution (4, 1.5 nm) (Figure 3a) and a random dispersion across the grid. Conversely, samples dispersed from cyclohexane in the presence of excess dodecanethiol showed a much narrower, single size distribution (3, 0.6 nm) and regions of highly ordered superlattice with close packing (Figure 3b), indicative of a surfactant effect.

IOP Conf. Series: Journal of Physics: Conf. Series **902** (2017) 012030 doi:10.1088/1742-6596/902/1/012030

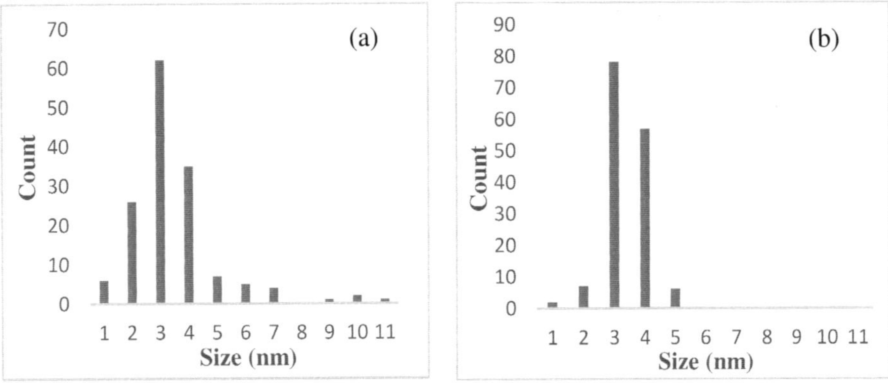

Figure 3. Particle size histograms of thiol stabilised Ag NP dispersed from (a) cyclohexane; and (b) cyclohexane in the presence of dodecanethiol, respectively.

Particle size histograms were generated using ImageJ, after magnification calibration, individual particles were measured (n = 150) and histograms plotted using Excel.

4. Discussion

The fundamental differences in dispersion of the Ag NP product may be explained in terms of the combined effects of the additional thiol surfactant on the Ag NP in colloidal suspension and solvent drying (convective deposition). It is considered that the Ag NP initial, mixed thiol stabilisation layer used for the synthesis is chemically bonded to the Ag NP and does not influence the processes of self-assembly or agglomeration, as observed here, and does not act as a surfactant in the same way as the dodecanethiol added to the cyclohexane, in advance of the deposition.

It has been reported that aggregate formation in solution may be prevented by the addition of small amounts of a stabilising agent to NP suspended in solution [4]. Accordingly, the self-assembly and separation of dispersed NP may be attributed to the presence of surfactant dodecanethiol molecules [5]. Attention should also be given to the additional 'coffee ring effect' consequent from the process of solvent evaporation [6] which may compete with the 2D superlattice formation. If the edge of the solvent is pinned to a surface, flow from the centre results as evaporation proceeds, with the accumulation of particles at the edge driven by capillary flow due to differential evaporation rates, competing with the induced Maragoni flow inside the droplet and the disruptive effect of the added surfactant which reduces the liquid surface tension gradient and further physically separates the dispersed NP upon self-assembly.

By way of illustration, Figures 4 depicts the self-assembly of close packed NP at the liquid-air interface with competing effects of convective flow and the surfactant effect of dodecanethiol addition, leading to very different interpretations of the Ag NP dispersion, once deposited on the TEM grid. In particular, the stabilising surfactant helps to maintain a mono-dispersed particle size in the solvent whilst promoting close packing of the NP on the TEM grid via attraction to the liquid-liquid interface [7]. Nucleation and subsequent development of superlattices occur during the drying process, enhanced by additional drying times associated with the presence of significant quantities of surfactant. In the absence of the stabilising surfactant, the Ag NP are capable of direct interaction leading to the observed variations in NP dispersion. Accordingly, care is needed when dispersing NP in the presence of a surfactant with regard to sample interpretation and the quantification of size and particle distribution.

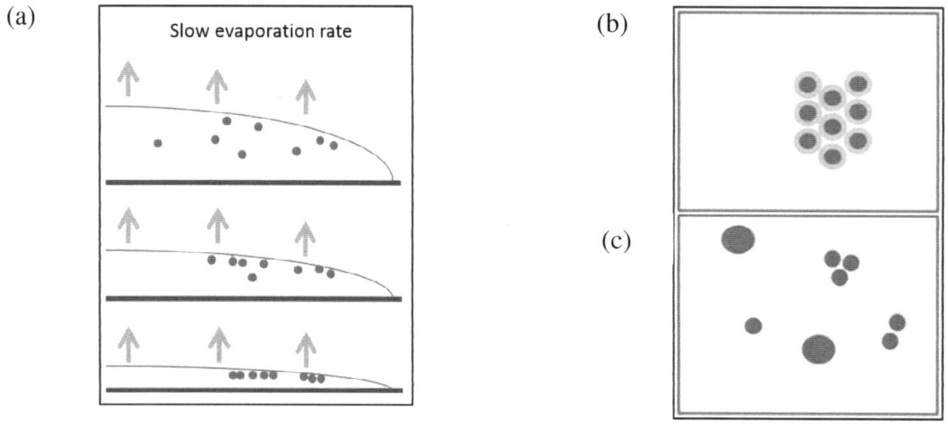

Figure 4. Schematic depiction of the drop casting process: and (a) the effect of dodecanethiol surfactant to NP in cyclohexane which facilitates partitioning of NP at the liquid-liquid interface and promotes NP superlattice formation. Schematic descriptions of NP product, as observed on the TEM grid: (b) Self assembled, mono-dispersed Ag NP clusters deposited from cyclohexane with excess dodecanethiol surfactant which acts to prevent particle agglomeration and promotes close packing superlattice formation; and (c) Bimodal agglomerates of Ag NP, as dispersed from cyclohexane.

5. Conclusions

The process of Ag NP dispersion is found to affect the apparent sample particle sizes and distributions. The addition of dodecanethiol to thiol stabilised Ag NP in cyclohexane leads to the deposition of self-assembled, mono-dispersed NP, whilst the absence of the stabilising surfactant appears to lead to the deposition of randomly distributed NP with a bimodal size range. The competing effects of convective deposition and surfactant effects are implicated in these different sample appearances.

Acknowledgements

The authors thank Whittards; Dr A. La Torre, the Nottingham Doctoral Training Centre (Complexity Science); the Engineering and Physical Sciences Research Council and Institute of Physics for support.

References

[1] Jiang C Cardin D J and Tsang S C 2008 Chemistry of Materials **20** (1): 14-16
[2] Carretero-Palacios S Jimenez-Solano A Miguez H 2016 ACS Energy Lett **1** (1): 323-331
[3] Brust M Walker M Bethell D Schiffrin D J and Whyman R 1994 Journal of the Chemical Society-Chemical Communications (7): 801-802
[4] Michen B Geers C Vanhecke D Endes C Rothen-Rutishauser B Balog S and Petri-Fink A 2015 Scientific Reports **5**: 9793
[5] Ma H and Hao J 2011 Chemical Society Reviews **40** (11): 5457-5471
[6] Deegan R D Bakajin O Dupont T F Huber G Nagel S R and Witten T A 1997 Nature **389** (6653): 827-829
[7] Bigioni T P Lin X M Nguyen T T Corwin E I Witten T A Jeager H M 2006 Nature Materials **5** 265-270

SEM tomography for the investigation of hybrid structures

**M Ferroni[1,2], M Donarelli[2], V Morandi[1], A Migliori[1], L Ortolani[1]
A Signoroni[2], C R Chandraiahgari[3,4], G De Bellis[3,4] and M S Sarto[3,4]**

[1] CNR – IMM National Research Council - Microelectronic and Microsystems Institute, via Gobetti 101, 40129 Bologna, Italy

[2] Department of Information Engineering, University of Brescia, Via Branze 38, 25123 Brescia, Italy

[3] Department of Astronautical, Electrical and Energy Engineering, Sapienza University of Rome, via Eudossiana 18, Rome 00184, Italy.

[4] Research Center for Nanotechnology Applied to Engineering of Sapienza (CNIS), SNNLab, Sapienza University of Rome, Piazzale Aldo Moro, 5, Rome 00185, Italy

E-mail: matteo.ferroni@unibs.it

Abstract. The morphological investigation at the micrometric scale of a graphene - ZnO nanorods hybrid structure is performed by scanning electron microscopy. When operated in the scanning-transmission imaging mode, the detection strategy allows implementation of a tomographic approach to recover the three dimensional spatial arrangement of the sample constituents. This tomographic approach complements the serial-sectioning imaging methods and is suitable for thin, self-standing specimens.

1. Introduction

Recent developments in material fabrication exploit the realization of hybrid structures combining different growth techniques and pursuing a positive tailoring of the functional properties.

As an example, graphene combined with ZnO crystalline nanorods (NRs) is presently under investigation because of the useful photocatalytic properties of this *n*-type semiconducting metal oxide and the electrical properties of graphene. In the form of crystalline NRs, ZnO has been used for a number of applications, ranging from gas sensors to solar cells [1-4]. A graphene-NRs system can be obtained by dispersing the NRs through the multilayered graphene flakes or promoting the nucleation of NRs directly over graphene [5-8]. In the present study, graphene flakes have been produced by solvo-thermal exfoliation method, and ZnO NRs have been grown in-situ through aqueous hydrothermal method [9]. A significant advantage of this approach is the compatibility with flexible and transparent polymeric substrates [10].

2. Characterization by scanning electron microscopy

The understanding and modelling of hybrid structures calls for an investigation of the spatial arrangement of the constituents at the micrometric scale. This intermediate magnification range could be easily covered by scanning electron microscopy (SEM); however, the interpretation of sample morphology from the conventional secondary-electron (SE) images should be carried out with caution, and quantitative three dimensional information is beyond the limits of the technique.

2.1. Conventional imaging with secondary electrons

Figure 1 (left and right) compares two low-magnification SEM images obtained with SE of a flake of graphene in-situ decorated with ZnO NRs and nanoparticles.

For the observation, the specimen was drop cast from the aqueous suspension over a carbon grid, according to a standard procedure for transmission electron microscopy (TEM) samples. When implemented in the SEM, the suspension of the thin sample over a lacy film prevents the saturation of the SE/BSE signals from the bulk substrate and, also, allows the exploitation of the transmitted-electrons imaging mode.

The two images in Figure 1, obtained with the standard Everhard-Torley (E-T) detector (left) and the In-lens detector (right) integrated in the Gemini column of the SEM (ZEISS-LEO 1530 FEG SEM), are quite different because of the limited difference in SE emission between the carbon-based material and nanorods dispersed at the top surface. The resulting images feature low contrast and a significant compositional contribution: in particular, the sensitivity of the E-T detector to the SE2 component of the SE signal, together with the +300 V bias of the collector grid, allows a significant compositional contribution related to the back-scattered electrons and promotes the visibility of most of the ZnO nanostructures underlying the graphene layer present at the centre of the selected area.

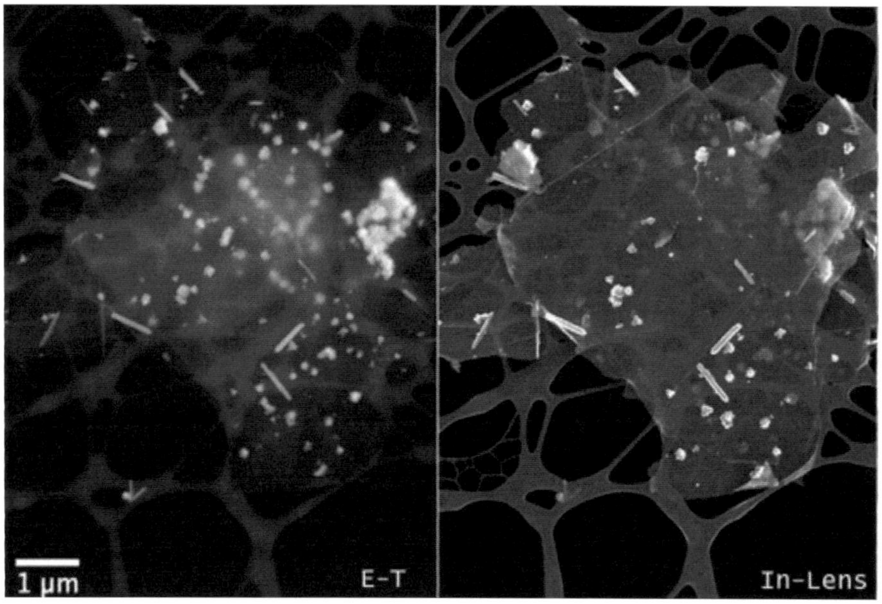

Figure 1: SEM images of a graphene – ZnO nanorods hybrid structure by secondary-electron imaging: Everhart-Thorley *(left)* and In Lens detector *(right)*.

2.2. Imaging with transmitted electrons

The scanning-transmission imaging mode in the SEM (STEM in SEM or T-SEM) is a standard implementation for the SEM platform, which has gained interest for the observation of light-element and nanostructured materials as a low-energy counterpart of TEM.

Figure 2 compares the TEM images (FEI Tecnai F20 microscope) of the graphene-ZnO sample obtained with conventional Bright-Field (left) and STEM-HAADF (center) with the one obtained by the STEM in SEM (right). Figure 2-right shows the transmitted-electron image of the very same area shown in Figure 1, obtained in the STEM-in-SEM, operated in a Bright-Field condition. In this image, the ZnO NRs and nanoparticles are clearly recognizable from graphene and the supporting lacy carbon film of the TEM grid.

The image contrast is monotonically governed by the sample density and projected thickness, while the resolution is limited by the size of the focused electron beam. This imaging mode meets the requirements for a tomographic approach to the three dimensional reconstruction of the mass distribution of the sample.

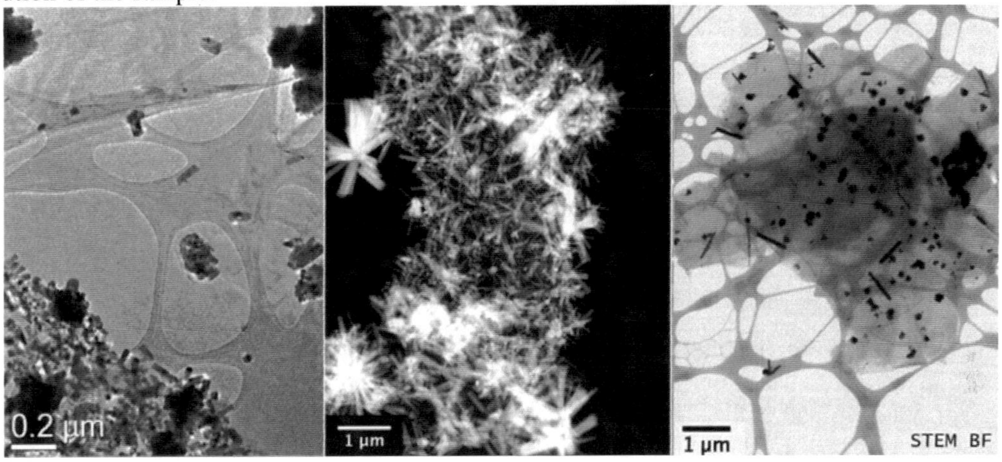

Figure 2: TEM and SEM images of the graphene – ZnO nanorods structure by transmitted-electron imaging. Conventional TEM Bright-Field *(left)* and TEM STEM-HAADF *(center)*. *(right)* SEM image in STEM Bright-Field condition.

3. Electron tomography in the SEM

The SEM platform is already used in the 3D visualization of samples in both biological and physical science through operation of the microscope in a *slice and view* mode, where a section of the bulk sample is progressively obtained by a focused ion beam and visualized with the electron beam. This approach performs well for uniform and compact specimens, as the section cutting is required to proceed evenly throughout the entire reconstructed volume. Limitations may also arise when the sample features a significant degree of porosity and voids. In the case of the graphene-ZnO NRs sample, such structure is not suitable for serial sectioning as the sample is extremely light, self-standing and lacks rigid support from an embedding media.

As a complementary approach to three dimensional imaging, the scanning-transmission imaging was recently integrated with a rotating sample holder to demonstrate a technically viable realization of electron tomography in the SEM [11-12]. The tomographic reconstruction algorithm introduces the additional requirement for the detection strategy: the STEM signal is based on the intensity of the electrons forward-scattered by the thin specimen, and the mass-thickness contrast should vary monotonically when the sample is rotated and projected from different directions.

For the graphene-ZnO NRs sample, the optimal imaging condition corresponds to the Dark-Field (DF) mode, owing to the difference in forward scattering between the thin graphene and the nanometric ZnO crystals. The image presented in Figure 3-left has been acquired in DF condition, so the highest signal-to-background ratio can be measured, the brightest areas being related to the ZnO nanorods. The light-grey area is the graphene flake, which is rippling over the carbon grid. The beam acceleration voltage was 27 kV, and the detection strategy was adapted to maintain a complete collection of the transmitted signal upon tilting the sample for the series of projections.

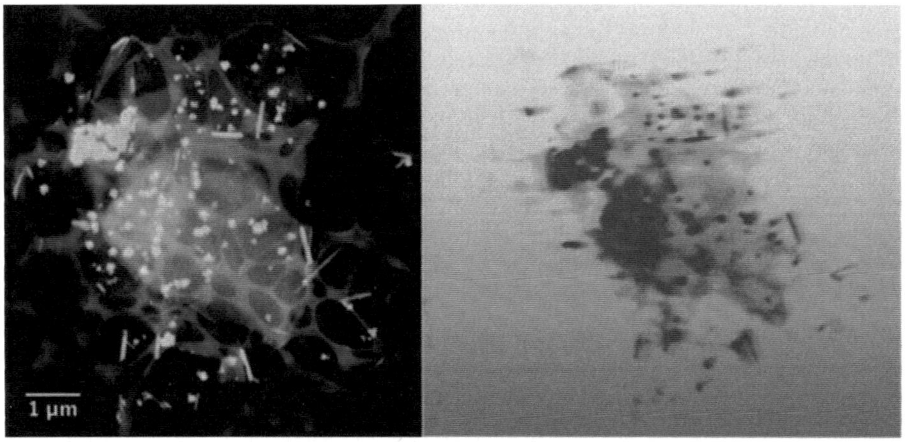

Figure 3: *(left)* STEM image of the graphene-NRs in dark-field condition.
(right) A colored rendering of the reconstructed tomogram.

The series of the sample DF projection images has been acquired by tilting the specimen from -60° to +50°, at 2° steps. The projection images have been aligned each other with a cross-correlation algorithm, and the result of the reconstruction of the 56 projection images is reported in Figure 3-right.

It turns out that the tomogram, obtained by standard retroprojection and followed by iterative refining, retrieves the correct spatial disposition of the hybrid system.

4. Discussion and concluding remarks

SEM is a valuable tool for the preliminary characterization and for the optimization of the preparation process of hybrid structures. In addition, the proposed approach to electron tomography in SEM is suitable for the assessment of the dispersion of ZnO NRs in graphene flakes. This method is potentially applicable to a large category of partially electron-transparent samples, and volumes up to 1 μm^3 could be reconstructed. The detection strategy could also take advantage of recent algorithmic implementations, such as discrete tomography or compressed sensing, to optimize the projection scheme and to reduce the number of projections required to retrieve the structure of the sample with nanometric resolution.

Acknowledgments

Ferroni M and Donarelli M acknowledge the financial support from the *Ministero degli Affari Esteri e della Cooperazione Internazionale* within the ITA-EGYPT CYAMOXSOLAR project.

References

[1] Wang H T, Kang B S and Ren F 2005 *Appl. Phys. Lett.* **86** 243503
[2] Jiaqiang X *et al* 2006 *Sens. and Act. B* **113** 526
[3] Gao T and Wang T H 2005 *Appl. Phys. A* **80** 1451
[4] Hsu Y F *et al* 2008 *Appl. Phys. Lett.* **92** 133507
[5] Zou R *et al* 2013 *J. Mater. Chem. A* **1** 8445
[6] Chandraiahgari C R *et al* 2015 *RSC Adv.* **5** 49861
[7] Ding J *et al* 2015 *RSC Adv.* **5** 22935
[8] Chung R *et al* 2013 *Nanoscale Res. Lett.* **8** 350
[9] Chandraiahgari C R *et al* 2016 *RSC Adv.* **6** 83217
[10] Casiraghi C and Whiters F 2016 in *2D Materials for Nanoelectronics* (Houssa M, Dimoulas A and Molle A, eds.) p 434
[11] Ferroni M *et al* 2016 *Scientific Reports* **6** 33354
[12] Ferroni M *et al* 2015 *J. Phys.: Conf. Ser.* **644** 012012

Armed conflict impacts on the microscale

Mol, L[1]

[1]Department of Geography and Environmental Management, University of the West of England, Bristol, BS16 1QY

Email: lisa.mol@uwe.ac.uk

Abstract. Armed conflict has left noticeable marks on our heritage, and has increasingly become a target in itself despite protective legislature such as the 1950 The Hague convention. Both built heritage and stone objects are partially destroyed through the impact of projectiles such as bullets and shrapnel. This study investigates alterations to the fabric of sandstone, a commonly used material for building and sculptures, upon impact using a combination of microscopy and SEM-EDX.

1. Introduction

Although by no means a new phenomenon, incidents of deliberate targeting of heritage during conflict have increased in frequency over the past three years, coinciding with the rise of IS / Daesh. Deliberate destruction of stone icons during the War of the Three Kingdoms (1640s), the destruction of Old Jerusalem (1948) and the Buddhas of Bamiyan (2001) are but a few examples of the scars left by ideologically-driven conflict on our heritage. What is of increasing concern is the advances over the past 50 years in the strength of both the weaponry and the ammunition [1] which has led to an ever-growing array of assault rifles and automatic guns weapons that can create substantial damage to any heritage unfortunate enough to be caught in cross-fire. While little can be done by the material science community to prevent the use of such weaponry, we can 'prepare for peace' by investigating the damage likely to occur to heritage and its implications for restoration post-conflict.

Of even greater concern is the very limited availability of any scientific knowledge that could aid in facilitating an understanding of the long-term consequences of these impacts on heritage conservation, and development of strategies needed to salvage conserve damaged heritage where possible. Where materials such as sand and metals have been investigated successfully [2,3], few attempts have been made to map the impact of ballistics into commonly used building materials such as the sandstone examined in this study. In particular, establishing the damage mechanisms behind a bullet impact is keystone knowledge in determining immediate triage as well as long-term conservation strategies [4]; using high-resolution methods such as Scanning Electron Microscopy (SEM) and optical microscopy this study investigates the physical effects of bullet impacts on sandstone surfaces and subsurface matrix. While the study presented here is limited to relatively small (.22 calibre lead bullet) ammunition, the measurements show that damage to the stone structure altered responses to moisture and temperature fluctuations.

2. Methodology

2.1 Materials used

The sandstone used in this study was quarried in the Huesca region in northeast Spain. This type of sandstone is well-consolidated mesoporous sandstone, (average pore size between 40 and 70μm; water absorption capacity of 1.8%). Test blocks measured 15 x 15 x 7.5 cm, with the larger 15 x 15cm surface used as the target. This lithology is particularly suitable for impact testing due to the homogenous matrix, which minimises the potential for test anomalies caused by variations in the sandstone. The samples were shot with .22 calibre lead bullets from a distance of 200m (figure 1). Six sandstone samples were impacted and two were used as control samples.

2.2 SEM/EDX and Optical Microscopy

SEM analysis was carried out to investigate lead deposition using an ESEM FEG. The SEM analysis was completed using a FEI Philips XL30 Environmental Scanning Electron Microscope Field Emission Gun (ESEM FEG) equipped with OSIRIS/ISIS data analysis software. Thin section imaging was undertaken with a Nikon Optiphot-pol microscope (with transmitted and incident halogen light illumination). The following objectives were used for transmitted light observations: x4/0.1 160/-; x10/0.25 160/-; x40/0.65 160/0.17. M-plan Differential Interference Contrast objectives were used for incident light observations: x5 0.1 210/0; x10/0.25 210/0; x40 0.65 210). To produce the images we used a Q-imaging (QICAM fast 1394) high performance IEEE FireWire 12-bit digital CCD (1392 x 1040, 1.4 million pixels, with a pixel size of 4.65μm x 4.65μm) camera system attached by a ½" C-mount. Live image capturing and processing was achieved using Syncroscopy's AcQuis (v. 4.0.1.8) software.

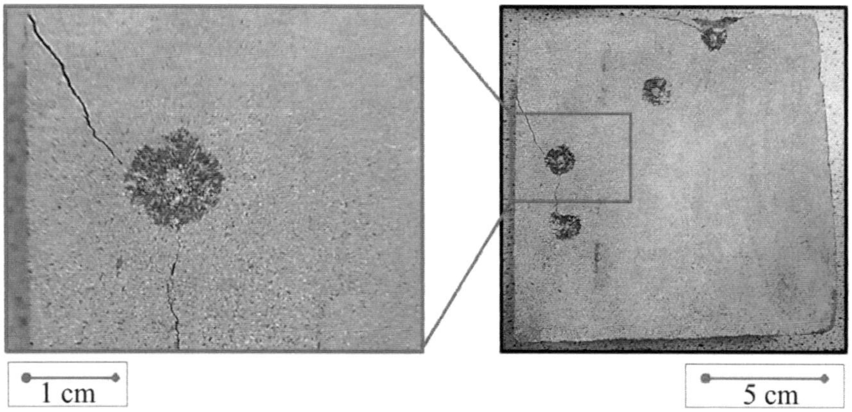

Figure 1: Optical images of a sample with .22 calibre lead bullet impacts

3. Results and discussion

3.1 Damage mapping

SEM-EDX analysis of the impact area showed noticeable deformation of the subsurface matrix of the sandstone, as well as loss of material from the surface. As figure 2[A] illustrates, the sandstone surface was disrupted and a friable indentation is left within the direct impact site. While the surface loss appeared to be minimal at first visual inspection (figure 1), the SEM-EDX analysis at 500μm scale showed that the material removal area extends sufficiently deep into the sub-surface to create a crater-like indentation. Further exacerbation of this damage is seen in figure 2[B]; compression of the subsurface at the point of impact in the sub-surface zone is resultant in a complex fracture network. The fracture network running parallel to the surface indicate an initial compression at the time of impact, followed by a release of compressive stress and the creation of parallel V-shaped fractures. The combination of this damage indicates that two damage mechanisms in bone which are differentiated in medical studies (crushing and stretching [5]) are simultaneously observed in sandstone impact sites, in the form of a friable surface within the direct impact zone (crushing) and the fracture network surrounding the impact zone (stretching).

The damage to the stone matrix is not limited to the larger stone structure; at still greater magnification, the damage to individual quartz grains is apparent both in the density of the fracturing within the grain, but also the shape of the selective removal of the quartz material. Figure 2[C] shows a quartz grain adjacent to the impact site which has been deformed into a crate-like formation, and is substantially fractured throughout. The surface of grain, when seen under higher magnification (figure 2[D]) appears to have been strained during compression to form plate structures parallel to the surface which are dislodged during the stress-release stage post impact. These two stages, compression and stress-release, mirror those observed at stone structure scale and show that bullet impacts result in the formation of a complex fracture network not only around the surface of the impact zone (as shown in figure 1) but also deeper within the subsurface.

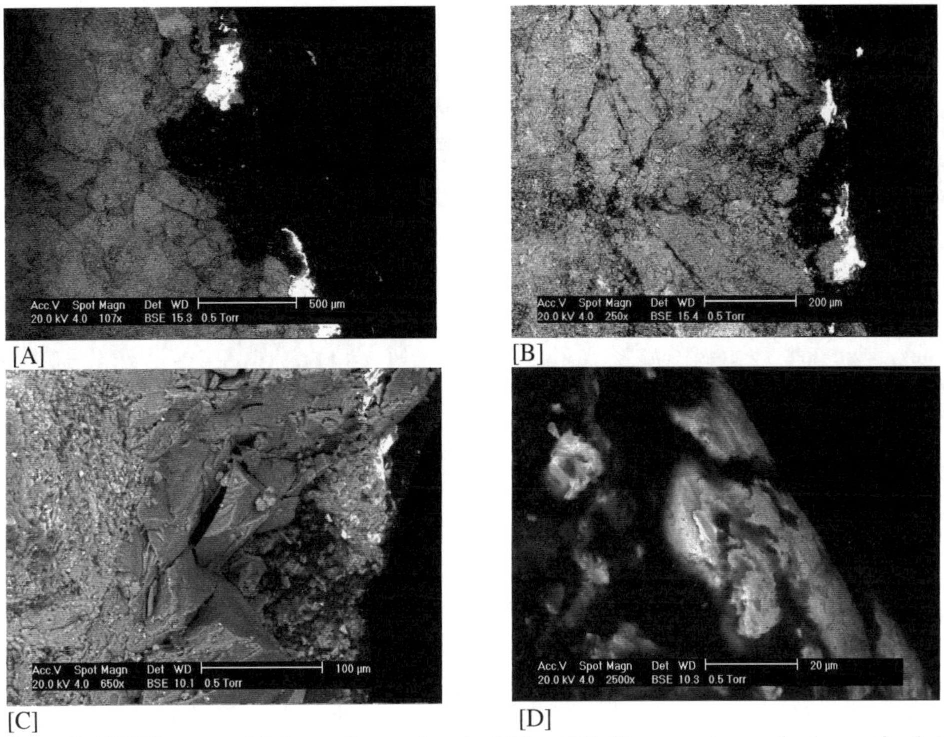

Figure 2: SEM images of deformation and material loss [A], Compression and release of sub-impact zone [B], localised damage to individual grains [C], fracturing of grain surface [D]

3.2 Ingress of lead into matrix

The final noticeable difference post-impact is the deposition of metals (lead) on and within the surface. The bullet used were strong enough to damage the surface but weak enough to deform and/or melt upon impact, leaving behind a thin film of lead in some areas. Figure 3 shows an overview of the impact area of sample 7 where clear lead residues were observed on the surface. These residues become more obvious when using plane polarised light (see figure 3[B]); the black areas indicate lead deposits. The lead has not only deposited on the surface (as shown on the right hand side of figure 3[B]) but also within the grains (middle and left). Figure 3[C] shows an enlargement of the middle section of 3[B] where the lead intrusion has propagated along a pre-existing fracture within in agrain and has fractured the grain, intruding not only through the centre but also into crevices extending from the centre fracture.

Further examination of the lead intrusion with SEM-EDX showed that the lead forms a thin film over the freshly exposed quartz grain surfaces on the impact site, as illustrated by figures 4[A] and 4[B]. This lead residue, identified using OSIRIS/ISIS software, could potentially affect the chemical and physical properties of the surface post-impact, and needs to be considered when considering options for long-term conservation or restoration of materials affected by bullet impacts.

Figure 3: [A] Photomicrograph in crossed-polars of an impact area showing the general setting of the lead-impacted minerals [B] Plane polarised image of the impact area of sample 7, showing lead intrusions into the surface (circled) [C] Close up of the centre section of the lead intrusions into the surface

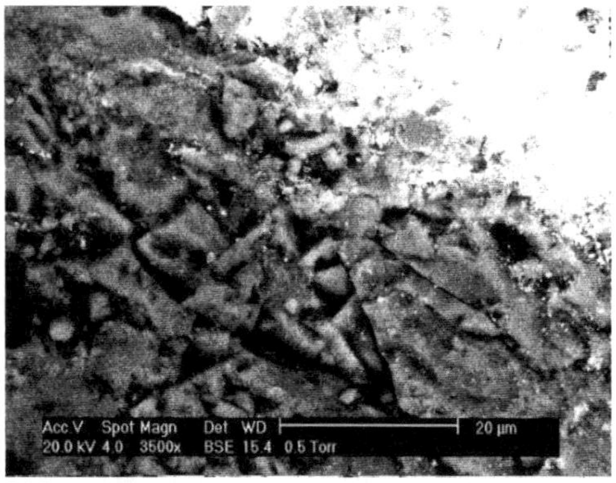

[A]

[B]

Figure 4: Illustration of thin lead film deposition

4. Conclusions

Though the study is focussed on a relatively small calibre bullet (.22), SEM and OM techniques have clearly demonstrated that bullet impacts on sandstone result in damage to the material at the macro- and micro-scales. This damage comprises material loss from the surface and the formation of a macro-scale fracture network in the subsurface. At the microscale, individual quartz grains are shown to have deformed and have lost material, presumably during stress release. Furthermore, this study has shown that lead form the bullet can penetrate and remain within the subsurface matrix and can even intrude and exploit pre-existing fractures in individual quartz grains. The combination of OM and SEM techniques is therefore a powerful tool for shedding new light on the impact of ballistics on stone surfaces, and could be considered a new and valuable approach to understanding the effects of conflict on built heritage. As shown in this study, impact damage reaches further into the subsurface than initially assumed with visual inspection, and therefore urgent work is needed to understand the impact of a wider array of weaponry ballistics on stone surfaces.

Acknowledgements

The author wishes to express her gratitude to Mr Owen Green for his valuable help and insights in optical microscopy and his assistance with the production of the thin sections. She also thanks the Herefordshire Breech-Loading Rifle Association for their assistance with the sample creations, and Dr Shaun Lavis for his helpful comments. Finally, she would like to thank Prof Heather Viles for the use of materials and equipment in the Oxford Rock Breakdown Laboratory.

References

[1] Lai, J., Guo, X., Zhu, Y. (2015) Repeated penetration and different depth explosion of ultra-high performance concrete. International Journal of Impact Engineering 84: 1 – 12

[2] Kiliç, N., Ekici, B. (2013) Ballistic resistance of high hardness armor steels against 7.62 mm armor piercing ammunition. Materials & Design 44: 35 - 48

[3] Walley, S.M. (2010) Historical review of high strain rate and shock properties of ceramics relevant to their application in armour. Advances in Applied Ceramics 109 (8) 446 – 466

[4] Mol, L., Gomez-Heras, M., Brassey, C., Green, O., Blenkinsop, T. (2017) The benefit of a tough skin: bullet impact, weathering and the preservation of heritage. Royal Society Open Science 4 160335

[5] Hollerman, J.J., et al (199) Gunshot wounds: 1. Bullets, ballistics and mechanism of injury. American Journal of Roentgenology 115: 685 – 690